Springer Series in Statistics

Advisors:
P. Bickel, P. Diggle, S. Fienberg, U. Gather,
I. Olkin, S. Zeger

Springer Series in Statistics

Alho/Spencer: Statistical Demography and Forecasting.
Andersen/Borgan/Gill/Keiding: Statistical Models Based on Counting Processes.
Atkinson/Riani: Robust Diagnostic Regression Analysis.
Atkinson/Riani/Ceriloi: Exploring Multivariate Data with the Forward Search.
Berger: Statistical Decision Theory and Bayesian Analysis, 2nd edition.
Borg/Groenen: Modern Multidimensional Scaling: Theory and Applications, 2nd edition.
Brockwell/Davis: Time Series: Theory and Methods, 2nd edition.
Bucklew: Introduction to Rare Event Simulation.
Cappé/Moulines/Rydén: Inference in Hidden Markov Models.
Chan/Tong: Chaos: A Statistical Perspective.
Chen/Shao/Ibrahim: Monte Carlo Methods in Bayesian Computation.
Coles: An Introduction to Statistical Modeling of Extreme Values.
Devroye/Lugosi: Combinatorial Methods in Density Estimation.
Diggle/Ribeiro: Model-based Geostatistics.
Efromovich: Nonparametric Curve Estimation: Methods, Theory, and Applications.
Eggermont/LaRiccia: Maximum Penalized Likelihood Estimation, Volume I: Density Estimation.
Fahrmeir/Tutz: Multivariate Statistical Modeling Based on Generalized Linear Models, 2nd edition.
Fan/Yao: Nonlinear Time Series: Nonparametric and Parametric Methods.
Ferraty/Vieu: Nonparametric Functional Data Analysis: Theory and Practice.
Fienberg/Hoaglin: Selected Papers of Frederick Mosteller.
Frühwirth-Schnatter: Finite Mixture and Markov Switching Models.
Ghosh/Ramamoorthi: Bayesian Nonparametrics.
Glaz/Naus/Wallenstein: Scan Statistics.
Good: Permutation Tests: Parametric and Bootstrap Tests of Hypotheses, 3rd edition.
Gouriéroux: ARCH Models and Financial Applications.
Gu: Smoothing Spline ANOVA Models.
Györfi/Kohler/Krzyżak/Walk: A Distribution-Free Theory of Nonparametric Regression.
Haberman: Advanced Statistics, Volume I: Description of Populations.
Hall: The Bootstrap and Edgeworth Expansion.
Härdle: Smoothing Techniques: With Implementation in S.
Harrell: Regression Modeling Strategies: With Applications to Linear Models, Logistic Regression, and Survival Analysis.
Hart: Nonparametric Smoothing and Lack-of-Fit Tests.
Hastie/Tibshirani/Friedman: The Elements of Statistical Learning: Data Mining, Inference, and Prediction.
Hedayat/Sloane/Stufken: Orthogonal Arrays: Theory and Applications.
Heyde: Quasi-Likelihood and its Application: A General Approach to Optimal Parameter Estimation.
Huet/Bouvier/Poursat/Jolivet: Statistical Tools for Nonlinear Regression: A Practical Guide with S-PLUS and R Examples, 2nd edition.
Ibrahim/Chen/Sinha: Bayesian Survival Analysis.
Jolliffe: Principal Component Analysis, 2nd edition.
Knottnerus: Sample Survey Theory: Some Pythagorean Perspectives.
Küchler/Sørensen: Exponential Families of Stochastic Processes.
Kutoyants: Statistical Inference for Ergodic Diffusion Processes.

(continued after index)

Peter J. Diggle
Paulo J. Ribeiro Jr.

Model-based Geostatistics

 Springer

Peter J. Diggle
Department of Mathematics and Statistics
Lancaster University, Lancaster, UK
LA1 4YF
p.diggle@lancaster.ac.uk

Paulo J. Ribeiro Jr.
Departmento de Estatística
Universidade Federal do Paraná
Curitiba, Paraná, Brazil
81.531-990
Paulojus@ufpr.br

Library of Congress Control Number: 2006927417

ISBN-10: 0-387-32907-2
ISBN-13: 978-0-387-32907-9

Printed on acid-free paper.

9 8 7 6 5 4 3 2 1

springer.com

For *Mandy, Silvia, Jono, Hannah, Paulo Neto and Luca*

Preface

Geostatistics refers to the sub-branch of spatial statistics in which the data consist of a finite sample of measured values relating to an underlying spatially continuous phenomenon. Examples include: heights above sea-level in a topographical survey; pollution measurements from a finite network of monitoring stations; determinations of soil properties from core samples; insect counts from traps at selected locations. The subject has an interesting history. Originally, the term *geostatistics* was coined by Georges Matheron and colleagues at Fontainebleau, France, to describe their work addressing problems of spatial prediction arising in the mining industry. See, for example, (Matheron, 1963; Matheron, 1971b). The ideas of the Fontainebleau school were developed largely independently of the mainstream of spatial statistics, with a distinctive terminology and style which tended to conceal the strong connections with parallel developments in spatial statistics. These parallel developments included work by Kolmogorov (1941), Matérn (1960, reprinted as Matérn, 1986), Whittle (1954, 1962, 1963), Bartlett (1964, 1967) and others. For example, the core geostatistical method known as *simple kriging* is equivalent to minimum mean square error prediction under a linear Gaussian model with known parameter values. Papers by Watson (1971, 1972) and the book by Ripley (1981) made this connection explicit. Cressie (1993) considered geostatistics to be one of three main branches of spatial statistics, the others being discrete spatial variation (covering distributions on lattices and Markov random fields) and spatial point processes. Geostatistical methods are now used in many areas of application, far beyond the mining context in which they were originally developed.

Despite this apparent integration with spatial statistics, much geostatistical practice still reflects its independent origins, and from a mainstream statistical perspective this has some undesirable consequences. In particular, explicit stochastic models are not always declared and *ad hoc* methods of inference are often used, rather than the likelihood-based methods of inference which are central to modern statistics. The potential advantages of using likelihood-based methods of inference are twofold: they generally lead to more efficient estimation of unknown model parameters; and they allow for the proper assessment of the uncertainty in spatial predictions, including an allowance for the effects of uncertainty in the estimation of model parameters.

Diggle, Tawn and Moyeed (1998) coined the phrase *model-based geostatistics* to describe an approach to geostatistical problems based on the application of formal statistical methods under an explicitly assumed stochastic model. This book takes the same point of view.

We aim to produce an applied statistical counterpart to Stein (1999), who gives a rigorous mathematical theory of kriging. Our intended readership includes postgraduate statistics students and scientific researchers whose work involves the analysis of geostatistical data. The necessary statistical background is summarised in an Appendix, and we give suggestions of further background reading for readers meeting this material for the first time.

Throughout the book, we illustrate the statistical methods by applying them in the analysis of real data-sets. Most of the data-sets which we use are publically available and can be obtained from the book's website, *http://www.maths.lancs.ac.uk/~diggle/mbg*.

Most of the book's chapters end with a section on computation, in which we show how the R software (R Development Core Team, 2005) and contributed packages **geoR** and **geoRglm** can be used to implement the geostatistical methods described in the corresponding chapters. This software is freely available from the R Project website (*http://www.r-project.org*).

The first two chapters of the book provide an introduction and overview. Chapters 3 and 4 then describe geostatistical models, whilst Chapters 5 to 8 cover associated methods of inference. The material is mostly presented for univariate problems i. e., those for which the measured response at any location consists of a single value but Chapter 3 includes a discussion of some multivariate extensions to geostatistical models and associated statistical methods.

The connections between classical and model-based gostatistics are closest when, in our terms, the assumed model is the linear Gaussian model. Readers who wish to confine their attention to this class of models on a first reading may skip Sections 3.11, 3.12, Chapter 4, Sections 5.5, 7.5, 7.6 and Chapter 8.

Many friends and colleagues have helped us in various ways: by improving our understanding of geostatistical theory and methods; by working with us on a range of collaborative projects; by allowing us to use their data-sets; and by offering constructive criticism of early drafts. We particularly wish to thank Ole Christensen, with whom we have enjoyed many helpful discussions. Ole is also the lead author of the **geoRglm** package.

Peter J. Diggle, Paulo J. Ribeiro Jr., March 2006

Contents

Preface		**v**
1	**Introduction**	**1**
	1.1 Motivating examples	1
	1.2 Terminology and notation	9
	1.2.1 Support	9
	1.2.2 Multivariate responses and explanatory variables . .	10
	1.2.3 Sampling design	12
	1.3 Scientific objectives	12
	1.4 Generalised linear geostatistical models	13
	1.5 What is in this book?	15
	1.5.1 Organisation of the book	16
	1.5.2 Statistical pre-requisites	17
	1.6 Computation	17
	1.6.1 Elevation data	17
	1.6.2 More on the `geodata` object	20
	1.6.3 Rongelap data	22
	1.6.4 The Gambia malaria data	24
	1.6.5 The soil data	24
	1.7 Exercises	26
2	**An overview of model-based geostatistics**	**27**
	2.1 Design	27
	2.2 Model formulation	28
	2.3 Exploratory data analysis	30
	2.3.1 Non-spatial exploratory analysis	30

    2.3.2   Spatial exploratory analysis   31
2.4   The distinction between parameter estimation and spatial
      prediction   35
2.5   Parameter estimation   36
2.6   Spatial prediction   37
2.7   Definitions of distance   39
2.8   Computation   40
2.9   Exercises   45

3 **Gaussian models for geostatistical data**   **46**
3.1   Covariance functions and the variogram   46
3.2   Regularisation   48
3.3   Continuity and differentiability of stochastic processes   . . .   49
3.4   Families of covariance functions and their properties   51
      3.4.1   The Matérn family   51
      3.4.2   The powered exponential family   53
      3.4.3   Other families   54
3.5   The nugget effect   56
3.6   Spatial trends   57
3.7   Directional effects   58
3.8   Transformed Gaussian models   60
3.9   Intrinsic models   63
3.10 Unconditional and conditional simulation   66
3.11 Low-rank models   68
3.12 Multivariate models   69
      3.12.1 Cross-covariance, cross-correlation and cross-variogram   70
      3.12.2 Bivariate signal and noise   71
      3.12.3 Some simple constructions   72
3.13 Computation   74
3.14 Exercises   77

4 **Generalized linear models for geostatistical data**   **79**
4.1   General formulation   79
4.2   The approximate covariance function and variogram   81
4.3   Examples of generalised linear geostatistical models   82
      4.3.1   The Poisson log-linear model   82
      4.3.2   The binomial logistic-linear model   83
      4.3.3   Spatial survival analysis   84
4.4   Point process models and geostatistics   86
      4.4.1   Cox processes   87
      4.4.2   Preferential sampling   89
4.5   Some examples of other model constructions   93
      4.5.1   Scan processes   93
      4.5.2   Random sets   94
4.6   Computation   94
      4.6.1   Simulating from the generalised linear model   94
      4.6.2   Preferential sampling   96

4.7   Exercises   97

5   Classical parameter estimation   **99**
5.1   Trend estimation   100
5.2   Variograms   100
    5.2.1   The theoretical variogram   100
    5.2.2   The empirical variogram   102
    5.2.3   Smoothing the empirical variogram   102
    5.2.4   Exploring directional effects   104
    5.2.5   The interplay between trend and covariance structure   105
5.3   Curve-fitting methods for estimating covariance structure . .   107
    5.3.1   Ordinary least squares   108
    5.3.2   Weighted least squares   108
    5.3.3   Comments on curve-fitting methods   110
5.4   Maximum likelihood estimation   112
    5.4.1   General ideas   112
    5.4.2   Gaussian models   112
    5.4.3   Profile likelihood   114
    5.4.4   Application to the surface elevation data   114
    5.4.5   Restricted maximum likelihood estimation for the
            Gaussian linear model   116
    5.4.6   Trans-Gaussian models   117
    5.4.7   Analysis of Swiss rainfall data   118
    5.4.8   Analysis of soil calcium data   121
5.5   Parameter estimation for generalized linear geostatistical
    models   123
    5.5.1   Monte Carlo maximum likelihood   124
    5.5.2   Hierarchical likelihood   125
    5.5.3   Generalized estimating equations   125
5.6   Computation   126
    5.6.1   Variogram calculations   126
    5.6.2   Parameter estimation   130
5.7   Exercises   132

6   Spatial prediction   **134**
6.1   Minimum mean square error prediction   134
6.2   Minimum mean square error prediction for the stationary
    Gaussian model   136
    6.2.1   Prediction of the signal at a point   136
    6.2.2   Simple and ordinary kriging   137
    6.2.3   Prediction of linear targets   138
    6.2.4   Prediction of non-linear targets   138
6.3   Prediction with a nugget effect   139
6.4   What does kriging actually do to the data?   140
    6.4.1   The prediction weights   141
    6.4.2   Varying the correlation parameter   144
    6.4.3   Varying the noise-to-signal ratio   146

6.5	Trans-Gaussian kriging	147
	6.5.1 Analysis of Swiss rainfall data (continued)	149
6.6	Kriging with non-constant mean	151
	6.6.1 Analysis of soil calcium data (continued)	151
6.7	Computation	151
6.8	Exercises	155

7   Bayesian inference   **157**

7.1	The Bayesian paradigm: a unified treatment of estimation and prediction	157
	7.1.1 Prediction using plug-in estimates	157
	7.1.2 Bayesian prediction	158
	7.1.3 Obstacles to practical Bayesian prediction	160
7.2	Bayesian estimation and prediction for the Gaussian linear model	160
	7.2.1 Estimation	161
	7.2.2 Prediction when correlation parameters are known .	163
	7.2.3 Uncertainty in the correlation parameters	164
	7.2.4 Prediction of targets which depend on both the signal and the spatial trend	165
7.3	Trans-Gaussian models	166
7.4	Case studies	167
	7.4.1 Surface elevations	167
	7.4.2 Analysis of Swiss rainfall data (continued)	169
7.5	Bayesian estimation and prediction for generalized linear geostatistical models	172
	7.5.1 Markov chain Monte Carlo	172
	7.5.2 Estimation	173
	7.5.3 Prediction	176
	7.5.4 Some possible improvements to the MCMC algorithm	177
7.6	Case studies in generalized linear geostatistical modelling . .	179
	7.6.1 Simulated data	179
	7.6.2 Rongelap island	181
	7.6.3 Childhood malaria in The Gambia	185
	7.6.4 *Loa loa* prevalence in equatorial Africa	187
7.7	Computation	193
	7.7.1 Gaussian models	193
	7.7.2 Non-Gaussian models	196
7.8	Exercises	196

8   Geostatistical design   **199**

8.1	Choosing the study region	201
8.2	Choosing the sample locations: uniform designs	202
8.3	Designing for efficient prediction	203
8.4	Designing for efficient parameter estimation	204
8.5	A Bayesian design criterion	206
	8.5.1 Retrospective design	206

        8.5.2   Prospective design   209
    8.6   Exercises   211

A Statistical background   **213**
    A.1   Statistical models   213
    A.2   Classical inference   213
    A.3   Bayesian inference   215
    A.4   Prediction   216

References   **218**

Index   **227**

1

Introduction

1.1 Motivating examples

The term *spatial statistics* is used to describe a wide range of statistical models and methods intended for the analysis of spatially referenced data. Cressie (1993) provides a general overview. Within spatial statistics, the term *geostatistics* refers to models and methods for data with the following characteristics. Firstly, values $Y_i : i = 1, \ldots, n$ are observed at a discrete set of sampling locations x_i within some spatial region A. Secondly, each observed value Y_i is either a direct measurement of, or is statistically related to, the value of an underlying continuous spatial phenomenon, $S(x)$, at the corresponding sampling location x_i. This rather abstract formulation can be translated to a variety of more tangible scientific settings, as the following examples demonstrate.

Example 1.1. *Surface elevations*

The data for this example are taken from Davis (1972). They give the measured surface elevations y_i at each of 52 locations x_i within a square, A, with side-length 6.7 units. The unit of distance is 50 feet ($\approx$15.24 meters), whereas one unit in y represents 10 feet ($\approx$3.05 meters) of elevation.

Figure 1.1 is a *circle plot* of the data. Each datum (x_i, y_i) is represented by a circle with centre at x_i and radius proportional to y_i. The observed elevations range between 690 and 960 units. For the plot, we have subtracted 600 from each observed elevation, to heighten the visual contrast between low and high values. Note in particular the cluster of low values near the top-centre of the plot.

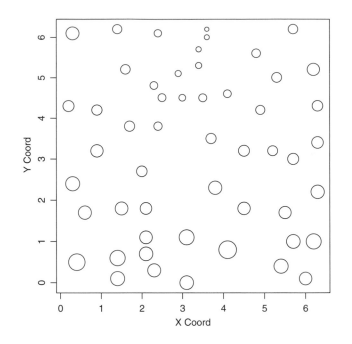

Figure 1.1. Circle plot of the surface elevation data. For the coordinates, the unit of distance is 50 feet. The observed elevations range from 690 to 960 units, where 1 unit represents 10 feet of elevation. Circles are plotted with centres at the sampling locations and radii determined by a linear transformation of the observed elevations (see Section 1.6).

The objective in analysing these data is to construct a continuous elevation map for the whole of the square region A. Let $S(x)$ denote the true elevation at an arbitrary location x. Since surface elevation can be measured with negligible error, in this example each y_i is approximately equal to $S(x_i)$. Hence, a reasonable requirement would be that the map resulting from the analysis should interpolate the data. Our notation, distinguishing between a measurement process Y and an underlying true surface S, is intended to emphasise that this is not always the case.

Example 1.2. *Residual contamination from nuclear weapons testing*

The data for this example were collected from Rongelap Island, the principal island of Rongelap Atoll in the South Pacific, which forms part of the Marshall Islands. The data were previously analysed in Diggle et al. (1998) and have the format $(x_i, y_i, t_i) : i = 1, \ldots, 157$, where x_i identifies a spatial location, y_i is a photon emission count attributable to radioactive caesium, and t_i is the time (in seconds) over which y_i was accumulated.

These data were collected as part of a more wide-ranging, multidisciplinary investigation into the extent of residual contamination from the U.S. nuclear weapons testing programme, which generated heavy fallout over the island in the 1950s. Rongelap island has been uninhabited since 1985, when the inhabitants left on their own initiative after years of mounting concern about the possible adverse health effects of the residual contamination. Each ratio y_i/t_i gives a crude estimate of the residual contamination at the corresponding location x_i but, in contrast to Example 1.1, these estimates are subject to non-negligible statistical error. For further discussion of the practical background to these data, see Diggle, Harper and Simon (1997).

Figure 1.2 gives a circle plot of the data, using as response variable at each sampling location x_i the observed emission count per unit time, y_i/t_i. Spatial coordinates are in metres, hence the east-west extent of the island is approximately 6.5 kilometres. The sampling design consists of a primary grid covering the island at a spacing of approximately 200 metres together with four secondary 5 by 5 sub-grids at a spacing of 50 metres. The role of the secondary sub-grids is to provide information about short-range spatial effects, which have an important bearing on the detailed specification and performance of spatial prediction methods.

The clustered nature of the sampling design makes it difficult to construct a circle plot of the complete data-set which is easily interpretable on the scale of the printed page. The inset to Figure 1.2 therefore gives an enlarged circle plot for the western extremity of the island. Note that the variability in the emission counts per unit time within each sub-grid is somewhat less than the overall variability across the whole island, which is as we would expect if the underlying variation in the levels of contamination is spatially structured.

In devising a statistical model for the data, we need to distinguish between two sources of variation: spatial variation in the underlying true contamination surface, $T(x)$ say; and statistical variation in the observed photon emission counts, y_i, given the surface $T(x)$. In particular, the physics of photon emissions suggests that a Poisson distribution would provide a reasonable model for the conditional distribution of each y_i given the corresponding value $T(x_i)$. The gamma camera which records the photon emissions integrates information over a circular area whose effective diameter is substantially smaller than the smallest distance (50 metres) between any two locations x_i. It is therefore reasonable to assume that the y_i are conditionally independent given the whole of the underlying surface $T(x)$. In contrast, there is no scientific theory to justify any specific model for $T(x)$, which represents the long-term cumulative effect of variation in the initial deposition, soil properties, human activity and a variety of natural environmental processes. We return to this point in Section 1.2.

One scientific objective in analysing the Rongelap data is to obtain an estimated map of residual contamination. However, in contrast to Example 1.1, we would argue that in this example the map should not interpolate the observed ratios y_i/t_i because each such ratio is a noisy estimate of the corresponding value of $T(x_i)$. Also, because of the health implications of the pattern of contamination across the island, particular properties of the map are of specific interest, for example the location and value of the maximum of $T(x)$, or areas within which $T(x)$ exceeds a prescribed threshold.

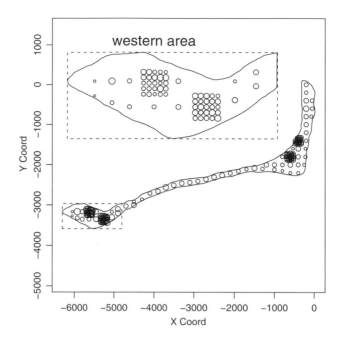

Figure 1.2. Circle plot for data from Rongelap island. Circles are plotted with centres at the sampling locations and radii proportional to observed emission counts per unit time. The unit of distance is 1 metre. The inset shows an enlargement of the western extremity of the island.

Example 1.3. *Childhood malaria in The Gambia*

These data are derived from a field survey into the prevalence of malaria para-sites in blood samples taken from children living in village communities in The Gambia, West Africa. For practical reasons, the sampled villages were concen-trated into five regions rather than being sampled uniformly across the whole country. Figure 1.3 is a map of The Gambia showing the locations of the sampled villages. The clustered nature of the sampling design is clear.

Within each village, a random sample of children was selected. For each child, a binary response was then obtained, indicating the presence or absence of malaria parasites in a blood sample. Covariate information on each child in-cluded their age, sex, an indication of whether they regularly slept under a mosquito net and, if so, whether or not the net was treated with insecticide. Information provided for each village, in addition to its geographical location, included a measure of the greenness of the surrounding vegetation derived from satellite data, and an indication of whether or not the village belonged to the primary health care structure of The Gambia Ministry for Health.

The data format for this example is therefore $(x_i, y_{ij}, d_i, d_{ij})$ where the subscripts i and j identify villages, and individual children within villages, re-spectively, whilst d_i and d_{ij} similarly represent explanatory variables recorded at the village level, and at the individual level, as described below. Note that if only village-level explanatory variables are used in the analysis, we might choose

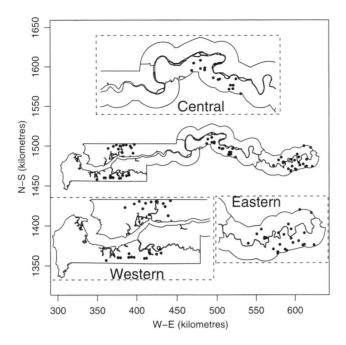

Figure 1.3. Sampling locations for The Gambia childhood malaria survey. The inset plots are enlarged maps of the western, central and eastern regions of The Gambia.

to analyse the data only at the village level, in which case the data format could be reduced to (x_i, n_i, y_i, d_i) where n_i is the number of children sampled in the ith village, and $y_i = \sum_{j=1}^{n_i} y_{ij}$ the number who test positive.

Figure 1.4 is a scatterplot of the observed prevalences, y_i/n_i, against the corresponding greenness values, u_i. This shows a weak positive correlation.

The primary objective in analysing these data is to develop a predictive model for variation in malarial prevalence as a function of the available explanatory variables. A natural starting point is therefore to fit a logistic regression model to the binary responses y_{ij}. However, in so doing we should take account of possible unexplained variation within or between villages. In particular, unexplained spatial variation between villages may give clues about as-yet unmeasured environmental risk factors for malarial infection.

Example 1.4. *Soil data*

These data have the format $(x_i, y_{i1}, y_{i2}, d_{i1}, d_{i2})$, where x_i identifies the location of a soil sample, the two y-variables give the calcium and magnesium content whilst the two d-covariates give the elevation and sub-area code of each sample.

The soil samples were taken from the 0-20 cm depth layer at each of 178 locations. Calcium and magnesium content were measured in $mmol_c/dm^3$ and the elevation in metres. The study region was divided into three sub-regions which have experienced different soil management regimes. The first, in the upper-left corner, is typically flooded during each rainy season and is no longer

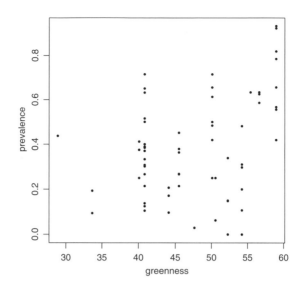

Figure 1.4. Observed prevalences against greenness for villages in The Gambia childhood malaria survey.

used as an experimental area because of its varying elevation. The calcium and magnesium levels in this region therefore represent the pattern of natural spatial variation in background content. The second, corresponding to the lower half of the study region, and the third, in the upper-right corner, have received fertilisers in the past: the second is typically occupied by rice fields, whilst the third is frequently used as an experimental area. Also, the second sub-region was the most recent of the three to which calcium was added to neutralise the effect of aluminium in the soil, which partially explains the generally higher measured calcium values within this sub-region.

The sampling design is an incomplete regular lattice at a spacing of approximately 50 metres. The data were collected by researchers from PESAGRO and EMBRAPA-Solos, Rio de Janeiro, Brasil (Capeche, 1997).

The two panels of Figure 1.5 show circle plots of the calcium (left panel) and magnesium (right panel) data separately, whilst Figure 1.6 shows a scatterplot of calcium against magnesium, ignoring the spatial dimension. This shows a moderate positive correlation between the two variables; the value of the sample correlation between the 178 values of calcium and magnesium content is $r = 0.33$.

Figure 1.7 shows the relationship between the potential covariates and the calcium content. There is a clear trend in the north-south direction, with generally higher values to the south. The relationships between calcium content and either east-west location or elevation are less clear. However, we have included on each of the three scatterplots a lowess smooth curve (Cleveland, 1981) which, in the case of elevation, suggests that there may be a relationship with calcium beyond an elevation threshold. Finally, the boxplots in the bottom right panel of Figure 1.7 suggest that the means of the distributions of calcium content are

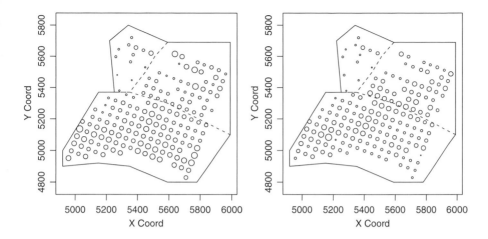

Figure 1.5. Circle plots of calcium (left panel) and magnesium (right panel) content with dashed lines delimiting sub-regions with different soil management practices.

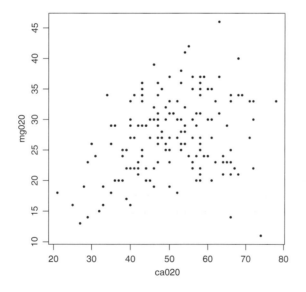

Figure 1.6. Scatterplot of calcium content against magnesium content in the 0-20 cm soil layer.

different in the different sub-regions. In any formal modelling of these data, it would also be sensible to examine covariate effects after allowing for a different mean response in each of the three sub-regions, in view of their different management histories.

One objective for these data is to construct maps of the spatial variation in calcium or magnesium content. Because these characteristics are determined from small soil cores, and repeated sampling at effectively the same location would yield different measurements, the constructed maps should not necessar-

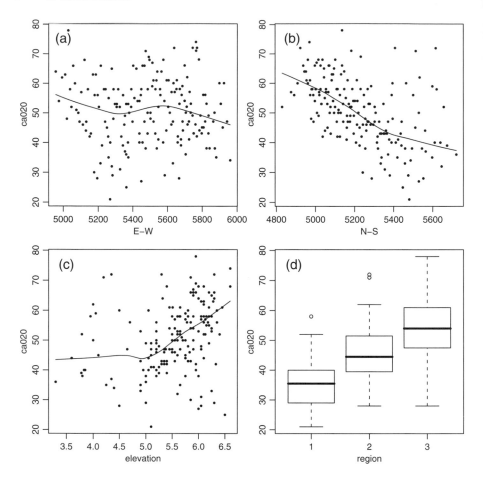

Figure 1.7. Scatterplots of calcium content against: (a) $E - W$ coordinate, (b) $N - S$ coordinate, (c) elevation. Lines are lowess curves. (d) Box-plots of calcium content in each of the three sub-regions.

ily interpolate the data. Another goal is to investigate relationships between calcium or magnesium content and the two covariates. The full data-set also includes the values of the calcium and magnesium content in the 20-40 cm depth layer.

We shall introduce additional examples in due course. However, these four are sufficient to motivate some basic terminology and notation, and to indicate the kinds of problems which geostatistical methods are intended to address.

1.2   Terminology and notation

The basic format for univariate *geostatistical data* is

$$(x_i, y_i) : i = 1, \ldots, n,$$

where x_i identifies a spatial location (typically in two-dimensional space, although one-dimensional and three-dimensional examples also occur) and y_i is a scalar value associated with the location x_i. We call y the *measurement variable* or *response*. A defining characteristic of geostatistics is that the measurement variable is, at least in principle, defined throughout a continuous study region, A say. Furthermore, we shall assume that the sampling design for the locations x_i is either deterministic (for example, the x_i may form a grid over the study region), or stochastically independent of the process which generates the measurements y_i. Each y_i is a realisation of a random variable Y_i whose distribution is dependent on the value at the location x_i of an underlying spatially continuous stochastic process $S(x)$ which is not directly observable. In particular cases, such as in our Example 1.1, we might reasonably assume that $Y_i = S(x_i)$, but in general it is important to preserve a distinction between the observable quantities Y_i and the unobservable, or latent process $S(x)$.

The basic form of a *geostatistical model* therefore incorporates at least two elements: a real-valued stochastic process $\{S(x) : x \in A\}$, which is typically considered to be a partial realisation of a stochastic process $\{S(x) : x \in \mathbb{R}^2\}$ on the whole plane; and a multivariate distribution for the random variable $Y = (Y_1, \ldots, Y_n)$ conditional on $S(\cdot)$. We call $S(x)$ the *signal* and Y_i the *response*. Often, Y_i can be thought of as a noisy version of $S(x_i)$ and the Y_i can be assumed to be conditionally independent given $S(\cdot)$.

1.2.1   Support

Examples 1.2 and 1.4 illustrate a general issue with geostatistical data concerning the *support* of each measured response. Formally, we associate each y_i with a point location x_i. However, in many cases y_i derives from a finite area for which x_i is a convenient reference point. In Example 1.4, the support is clearly identifiable as the circular cross-section of the soil core used to obtain each sample, and x_i denotes the centre of the cross-section. In Example 1.2, definition of the support is more difficult. The gamma camera integrates positron emissions over a circular neighbourhood of each sample location x_i, but rather than a sharp cut-off at a known distance, the camera traps a smaller proportion of the actual emissions with increasing distance from the centre of the circle. This implies that the modelled signal, $S(x)$, should strictly be interpreted as a weighted integral of an underlying spatially continuous signal, $S^*(x)$ say, so that

$$S(x) = \int w(r) S^*(x - r) dr.$$

Under this formulation, $S(x)$ is still a real-valued, spatially continuous process i.e., it is well-defined for all $x \in \mathbb{R}^2$. Its genesis as an integral does, however, have implications for what covariance structure we can strictly assume for the process $S(\cdot)$, since any smoothness in the behaviour of the weighting function $w(\cdot)$ constrains the allowable form of covariance structure for $S(\cdot)$. In this particular example we do not need to model the effect of the weighting function explicitly, because its effective range is much smaller than the minimum distance of 50 metres between any two points in the design.

The idea that geostatistical measurements have finite, rather than infinitesimal, support is to be contrasted with problems in which measurements are derived from a partition of a spatial region into discrete spatial units $i = 1, \ldots, n$, each of which yields a measurement y_i. This is often the case, for example, in spatial epidemiology, where data on disease prevalence may be recorded as counts in administrative sub-regions, for example counties or census tracts. In that context, the modelling options are either to deal explicitly with the effects of the spatial integration of an underlying spatially continuous process $S^*(x)$ or, more pragmatically, to specify a model at the level of the discrete spatial units i.e., a multivariate distribution for random variables $Y_i : i = 1, \ldots, n$. Models of the second kind have an extensive literature and are widely used in practice to analyse data arising as a result of spatial aggregation into discrete units. Less commonly, the actual spatial units are genuinely discrete; an example would be data on the yields of individual fruit trees in an orchard.

Evidently, a common feature of geostatistical models and discrete spatial models is that they both specify the joint distribution of a spatially referenced, n-dimensional random variable $(Y_1, \ldots, Y_n)$. An important difference is that a geostatistical model automatically embraces any n, and any associated set of sampling locations, whereas a discrete spatial model is specific to a particular set of locations. A classic early reference to the modelling and analysis of data from discrete spatial units is Besag (1974). See also Cressie (1993, chapters 6 and 7).

1.2.2   *Multivariate responses and explanatory variables*

As our motivating examples llustrate, in many applications the basic (x_i, y_i) format of geostatistical data will be extended in either or both of two ways. There may be more than one measurement variable, so defining *multivariate response*, $y_i = \{y_{i1}, \ldots, y_{id}\}$, or the data may include spatial explanatory variables, $\{d_k(x) : x \in A\}$, sometimes also called *covariates*.

The distinction between the two is not always clear-cut. From a modelling point of view, the difference is that a model for a multivariate response requires the specification of a vector-valued stochastic process over the study region A, whereas spatial explanatory variables are treated as deterministic quantities with no associated stochastic model. One consequence of this is that a spatial explanatory variable must, at least in principle, be available at any location within A if it is to be used to predict responses at unsampled locations x. An example would be the greenness index in Example 1.3. The index is calculated on a 1 km pixel grid and can therefore be used to predict malaria prevalence without making any assumptions about its spatial variation. Even then, in our experience the distinction between a stochastic signal $S(x)$ and a spatial explanatory variable $d(x)$ is largely a reflection of our scientific goals. Again using Example 1.3 to illustrate the point, the goal in this example is to understand how environmental factors affect malaria prevalence. Elevation is one of several factors which determine the suitability of a particular location to support breeding mosquitos, and is a candidate for inclusion as an explanatory variable in a stochastic model for prevalence. In contrast, in Example 1.1 the goal is to interpolate or smooth a spatially sparse set of measured elevations so as to obtain a spatially continuous elevation map, hence elevation is treated as a stochastic response.

In most geostatistical work, the adoption of a stochastic model for $S(x)$ reflects its unknown, unobserved quality rather than a literal belief that the underlying spatial surface of interest is generated by the laws of probability. Indeed, in many applications the role of the signal process $S(x)$ is as a surrogate for unmeasured explanatory variables which influence the response variable. In modelling $S(x)$ as a stochastic process we are using stochasticity at least in part as a metaphor for ignorance.

For this reason, when relevant explanatory variables are only available at the data locations x_i and we wish to use their observed values for spatial prediction at an unsampled location x, a pragmatic strategy is to treat such variables as additional responses, and accordingly to formulate a multivariate model. Example 1.4 illustrates both situations: the calcium and magnesium contents form a bivariate spatial stochastic process, whereas region and, to a good approximation, elevation, available at any location, are not of scientific interest in themselves, and can therefore be treated as explanatory variables. In this example, both components of the bivariate response are measured at each data location. More generally, measurements on different components of a multivariate response need not necessarily be made at a common set of locations.

Note that the locations x_i potentially play a dual role in geostatistical analysis. Firstly, spatial location is material to the model for the signal process $S(x)$ in that the stochastic dependence between $S(x)$ and $S(x')$ is typically modelled as a function of the locations in question, x and x'. Secondly, each location defines the values of a pair of explanatory variables corresponding to the two spatial coordinates. The convention in geostatistics is to use the term *trend surface*, to mean a spatially varying expectation of the response variable which is specified as a function of the coordinates of the x_i, whereas the term *external trend* refers to a spatially varying expectation specified as a function of other explanatory variables $d(x)$. For example, the elevation data as presented in Example 1.1 do not include any explanatory variables which could be used in an external trend model, but as we shall see in Chapter 2 a low-order polynomial trend surface can explain a substantial proportion of the observed spatial variation in the data.

1.2.3   Sampling design

The locations x_i at which measurements are made are collectively called the *sampling design* for the data. A design is *non-uniform* if the sampling intensity varies systematically over the study region, in the sense that before the actual sampling points are chosen, some parts of the study region are deliberately sampled more intensively than others. This is as distinct from the sampling intensity varying by chance; for example, if sample points are located as an independent random sample from a uniform distribution over the study region, it may (indeed, will) happen that some parts of the study region are more intensively sampled than others, but we would still describe this as a uniform design because of its method of construction.

A design is *non-preferential* if it is deterministic, or if it is stochastically independent of $S(\cdot)$. Conventional geostatistical methods assume, if only implicitly, that the sampling design is non-preferential, in which case we can legitimately analyse the data conditional on the design. Provided that the sampling process is non-preferential, the choice of design does not impact on the assumed model for the data, but does affect the precision of inferences which can be made from the data. Furthermore, different designs are efficient for different kinds of inference. For example, closely spaced pairs of sample locations are very useful for estimating model parameters, but would be wasteful for spatial prediction using a known model.

1.3   Scientific objectives

In most applications, the scientific objectives of a geostatistical analysis are broadly of two kinds: estimation and prediction.

Estimation refers to inference about the parameters of a stochastic model for the data. These may include parameters of direct scientific interest, for example those defining a regression relationship between a response and an explanatory variable, and parameters of indirect interest, for example those defining the covariance structure of a model for $S(x)$.

Prediction refers to inference about the realisation of the unobserved signal process $S(x)$. In applications, specific prediction objectives might include prediction of the realised value of $S(x)$ at an arbitrary location x within a region of interest, A, typically presented as a map of the predicted values of $S(x)$, or prediction of some property of the complete realisation of $S(x)$ which is of particular relevance to the problem in hand. For example, in the mining applications for which geostatistical methods were originally developed, the average value of $S(x)$ over an area potentially to be mined would be of direct economic interest, whereas in the Rongelap island example an identification of those parts of the island where $S(x)$ exceeds some critical value would be more useful than the average as an indicator of whether the island is fit for rehabilitation. Geostatistical models and methods are particularly suited to scientific problems whose objectives include prediction, in the sense defined here.

A third kind of inferential problem, namely *hypothesis testing*, can also arise in geostatistical problems, although often only in a secondary sense, for example in deciding whether or not to include a particular explanatory variable in a regression model. For the most part, in this book we will tacitly assume that testing is secondary in importance to estimation and prediction.

1.4 Generalised linear geostatistical models

Classical generalised linear models, introduced by Nelder and Wedderburn (1972), provide a unifying framework for the analysis of many superficially different kinds of independently replicated data. Several different ways to extend the generalised linear model class to dependent data have been proposed, amongst which perhaps the most widely used are *marginal models* (Liang and Zeger, 1986) and *mixed models* (Breslow and Clayton, 1993). What we shall call a *generalised linear geostatistical model* is a generalised linear mixed model of a form specifically oriented to geostatistical data.

The first ingredient in this class of models is a stationary Gaussian process $S(x)$. A stochastic process $S(x)$ is a *Gaussian model* if the joint distribution of $S(x_1), \ldots, S(x_n)$ is multivariate Gaussian for any integer n and set of locations x_i. The process is *stationary* if the expectation of $S(x)$ is the same for all x, the variance of $S(x)$ is the same for all x and the correlation between $S(x)$ and $S(x')$ depends only on $u = ||x - x'||$, the Euclidean distance between x and x'. We shall use the class of stationary Gaussian processes as a flexible, empirical model for an irregularly fluctuating, real-valued spatial surface. Typically, the nature of this surface, which we call the *signal*, is of scientific interest but the surface itself cannot be measured directly. The range of applicability of the model can be extended by the use of mathematical transformations. For example, in the suggested model for the Rongelap island photon emission data, the Gaussian process $S(x)$ is the logarithm of the underlying contamination surface $T(x)$. We discuss the Gaussian model, including non-stationary versions, in more detail in Chapter 3.

The second ingredient in the generalised linear geostatistical model is a statistical description of the data generating mechanism conditional on the signal. This part of the model follows a classical generalized linear model as described by McCullagh and Nelder (1989), with $S(x)$ as an offset in the linear predictor. Explicitly, conditional on $S(\cdot)$ the responses $Y_i : i = 1, \ldots, n$ at locations $x_i : i = 1, \ldots, n$ are mutually independent random variables whose conditional expectations, $\mu_i = \mathrm{E}[Y_i|S(\cdot)]$, are determined as

$$h(\mu_i) = S(x_i) + \sum_{k=1}^{p} \beta_k d_k(x_i), \qquad (1.1)$$

where $h(\cdot)$ is a known function, called the *link function*, the $d_k(\cdot)$ are observed *spatial explanatory variables* and the β_k are unknown *spatial regression parameters*. The terms on the right-hand side of (1.1) are collectively called the *linear predictor* of the model. The conditional distribution of each Y_i given $S(\cdot)$ is called the *error distribution*.

For each of our introductory examples, there is a natural candidate model within the generalized linear family.

For Example 1.1, in which the response is real-valued, we might adopt a *linear Gaussian* model, in which the link function $h(\cdot)$ is the identity and the error distribution is Gaussian with variance τ^2. Hence, the true surface elevation at a location x is given by $S(x)$ and, conditional on the realisation of $S(x)$ at all locations the measured elevations y_i are mutually independent, normally distributed with conditional means $S(x_i)$ and common conditional variance τ^2. A possible extension of this model would be to include spatial explanatory variables to account for a possible non-stationarity of $S(\cdot)$. For example, the circle plot of the data (Figure 1.1) suggests that elevations tend to decrease as we move from south to north. We might therefore consider including the north-south coordinate of the location as an explanatory variable, $d_1(\cdot)$ say, so defining a non-constant plane over the area. The conditional mean of each y_i given $S(x)$ would then be modelled as $d_1(x_i)\beta + S(x_i)$.

For Example 1.2, in which the response is a photon emission count, the underlying physics motivates the Poisson distribution as a suitable error distribution, whilst the log-linear formulation suggested earlier is an empirical device which constrains the expected count to be non-negative, as required. The photon emission counts Y_i can then be modelled as conditionally independent Poisson-distributed random variables, given an underlying surface $T(\cdot)$ of true levels of contamination. Also, the expectation of Y_i is directly proportional both to the value of $T(x_i)$ and to the time, t_i, over which the observed count is accumulated. Hence, the conditional distribution of Y_i should be Poisson with mean $t_i T(x_i)$. In the absence of additional scientific information a pragmatic model for $T(x)$, recognising that it necessarily takes non-negative values, might be that $\log T(x) = S(x)$ is a Gaussian stochastic process with mean μ, variance σ^2 and correlation function $\rho(x, x') = \mathrm{Corr}\{S(x), S(x')\}$. Like any statistical model, this is an idealisation. A possible refinement to the Poisson assumption for the emission counts conditional on the signal $S(x)$ would be to recognise that each y_i is a so-called nett count, calculated by subtracting from the raw count an estimate of that part of the count which is attributable to broad-band background radiation. With regard to the model for $S(x)$, the assumed constant mean could be replaced by a spatially varying mean if there were evidence of systematic variation in contamination across the island.

For Example 1.3, the sampling mechanism leads naturally to a binomial error distribution at the village level or, at the child level, a Bernoulli distribution with the conditional mean μ_{ij} representing the probability of a positive response from the jth child sampled within the ith village. A logit-linear model, $h(\mu_{ij}) = \log\{\mu_{ij}/(1 - \mu_{ij})\}$, constrains the μ_{ij} to lie between 0 and 1 as required, and is one of several standard choices. Others include the probit link, $h(\mu) = \Phi^{-1}(\mu)$ where $\Phi(\cdot)$ denotes the standard Gaussian distribution function, or the complementary-log-log, $h(\mu) = \log\{-\log(\mu)\}$. In practice, the logit and probit links are hard to distinguish, both corresponding to a symmetric

S-shaped curve for μ as a function of the linear predictor with the point of symmetry at $\mu = 0.5$, whereas the complementary-log-log has a qualitatively different, asymmetric form.

Example 1.4 features a bivariate response, and therefore falls outside the scope of the (univariate) generalized linear geostatistical model as described here. However, a separate linear Gaussian model could be used for each of the two responses, possibly after appropriate transformation, and dependence between the two response variables could then be introduced by extending the unobserved Gaussian process $S(x)$ to a bivariate Gaussian process, $S(x) = \{S_1(x), S_2(x)\}$. This example also includes explanatory variables as shown in Figure 1.7. These could be added to the model as indicated in equation (1.1), using the identity link function.

1.5 What is in this book?

This books aims to describe and explain statistical methods for analysing geostatistical data. The approach taken is model-based, by which we mean that the statistical methods are derived by applying general principles of statistical inference based on an explicitly declared stochastic model of the data generating mechanism.

In principle, we place no further restriction on the kind of stochastic model to be specified. Our view is that a model for each particular application should ideally be constructed by collaboration between statistician and subject-matter scientist with the aim that the model should incorporate relevant contextual knowledge whilst simultaneously avoiding unnecessary over-elaboration and providing an acceptable fit to the observed data. In practice, a very useful and flexible model class is the generalized linear geostatistical model, which we described briefly in Section 1.4. Chapters 3 and 4 develop linear and generalized linear geostatistical models in more detail. We also include in Chapter 4 some cautionary examples of spatial modelling problems for which the generalized linear model is inadequate.

We shall develop both classical and Bayesian approaches to parameter estimation. The important common feature of the two approaches is that they are based on the likelihood function. However, we also describe simpler, more ad hoc approaches and indicate why they are sometimes useful.

For problems involving prediction, we shall argue that a Bayesian approach is natural and convenient because it provides a ready means of allowing uncertainty in model parameters to be reflected in the widths of our prediction intervals.

Within the Bayesian paradigm, there is no formal distinction between an unobserved spatial stochastic process $S(x)$ and an unknown parameter θ. Both are modelled as random variables. Nevertheless, although we use Bayesian methods extensively, we think that maintaining the distinction between *prediction* of $S(x)$ and *estimation* of θ is important in practice. As noted in Section 1.3 above, prediction is concerned with learning about the particular realisation of the stochastic process $S(x)$ which is assumed to have generated the observed data y_i, whereas estimation is concerned with *properties* of the process $S(\cdot)$ which apply to all realisations. Section 2.4 discusses some of the inferential implications of this distinction in the context of a specific, albeit hypothetical, example.

1.5.1   Organisation of the book

Chapters 3 and 4 of the book discuss geostatistical models, whilst Chapters 5 to 8 discuss associated methods for the analysis of geostatistical data. Embedded within these chapters is a model-based counterpart to classical, linear geostatistics, in which we assume that the linear Gaussian model is applicable, perhaps after transformation of the response variable. We do not necessarily believe that the Gaussian is a correct model, only that it provides a reasonable approximation. Operationally, its significance is that it gives a theoretical justification for using linear prediction methods, which under the Gaussian assumption have the property that they minimise mean squared prediction errors. In Chapter 8 we give a model-based perspective on design issues for geostatistical studies.

Our aim has been to give a thorough description of core topics in model-based geostatistics. However, in several places we have included shorter descriptions of some additional topics, together with suggestions for further reading. These additional topics are ones for which model-based geostatistical methods are, at the time of writing, incompletely developed. They include constructions for multivariate Gaussian models, preferential sampling and point process models.

Throughout the book, we intersperse methodological discussion with illustrative examples using real or simulated data. Some of the data-sets which we use are not freely available. Those which are can be downloaded from the book's website, *http://www.maths.lancs.ac.uk/~diggle/mbg*.

Most chapters, including this one, end with a section on "Computation." In each such section we give examples of R code to implement the geostatistical methods described in the corresponding chapters, and illustrate some of the optional input parameters for various functions within the contributed R packages **geoR** and **geoRglm**. These illustrations are intended to be less formal in style than the help pages which form part of the package documentation. The websites, *http://www.est.ufpr.br/geoR* and *http://www.est.ufpr.br/geoRglm*, also include illustrative sessions using these two packages. Material from the computation sections is also available from the book's website.

The "Computation" sections assume that the reader is familiar with using R for elementary statistics and graphics. For readers who are not so familiar, a good introductory textbook is Dalgaard (2002), whilst general information about the R project can be found in documentation available in the R-Project website, *http://www.r-project.org*. These sections are also optional, in the sense that they introduce no new statistical ideas, and the remainder of the book can be read without reference to this material.

1.5.2   Statistical pre-requisites

We assume that the reader has a general knowledge of the standard tools for exploratory data analysis, regression modelling and statistical inference. With regard to regression modelling, we use both linear and generalised linear models. One of many good introductions to linear models is Draper and Smith (1981). The standard reference to generalised linear models is McCullagh and Nelder (1989). We make extensive use of likelihood-based methods, for both non-Bayesian and Bayesian inference. The Appendix gives a short summary of the key ideas. A good treatment of likelihood-based methods in general is Pawitan (2001), whilst O'Hagan (1994) specifically discusses the Bayesian method.

Readers will also need some knowledge of elementary probability and stochastic process theory. Introductory books at a suitable level include Ross (1976) for elementary probability and Cox and Miller (1965) for stochastic processes.

We shall also use a variety of computer-intensive methods, both for simulating realisations of stochastic processes and more generally in Monte Carlo methods of inference, including Markov chain Monte Carlo. A good general introduction to simulation methods is Ripley (1987). Tanner (1996) presents a range of computational algorithms for likelihood-based and Bayesian inference. Gelman, Carlin, Stern and Rubin (2003) focus on Bayesian methods for a range of statistical models. Gilks, Richardson and Spiegelhalter (1996) discuss both theoretical and practical aspects of Markov chain Monte Carlo.

1.6   Computation

The examples in this section, and in later chapters, use the freely available software R and the contributed R packages **geoR** and **geoRglm**. Readers should consult the R project website, *http://www.r-project.org*, for further information on the software and instructions on its installation.

In the listing of the R code for the examples, the > sign is the R prompt and the remainder of the line denotes the R command entered by the user in response to the prompt. R commands are shown in `slanted verbatim font like this`. When a single command is spread over two or more lines, the second and subsequent lines of input are prompted by a + sign, rather than the > sign. The R system is based on subroutines called *functions*, which in turn can take *arguments* which control their behaviour. Function names are followed by parentheses, in the format `function()`, whereas arguments are written within the parentheses. Any lines without the > prompt represent outputs from a function which, by default, are passed back to the screen. They are shown in `verbatim font like this`.

1.6.1   Elevation data

In our first example, we give the commands needed to load the **geoR** package, and to produce the circle plot of the elevation data, as shown in Figure 1.1.

The example assumes that the data are stored in a standard three-column text-file `elevation.dat` located in the R working directory. The first two columns on each line give the (x, y)-coordinates of a location, whilst the third column gives the corresponding value of the measured elevation. The version of the data which can be downloaded from the book website is already formatted in this way.

```
> require(geoR)
> elevation <- read.geodata("elevation.dat")
> points(elevation, cex.min = 1, cex.max = 4)
```

The first command above uses the built-in R function `require()` to load the **geoR** package. The second command reads the data and converts them to an object of the class `geodata` using `read.table()` and `as.geodata()` internally. The last command invokes a method for `points()` which is provided by the package **geoR**. In this way, the generic R function `points()` is able to use the **geoR** function `points.geodata()` to produce the required plot of the data. The example includes optional settings for arguments which control the sizes of the plotted circles. By default, the diameters of the plotted circles are defined by a linear transformation of the measured elevations onto a scale ranging between `cex.min` and `cex.max` times the default plotting character size.

The output returned when typing `args(points.geodata)` will show other arguments which can be used to modify the resulting plot. For example,

```
> points(elevation, cex.min = 2, cex.max = 2, col = "gray")
```

will plot the locations as filled circles with grey shades proportional to the measured elevation values, whereas

```
> points(elevation, cex.min = 2, cex.max = 2, pt.div = "quint")
```

will result in points filled with different colours according to the quintiles of the empirical distribution of measured elevations.

Because the elevation data are also included in the **geoR** package, they can be loaded from within R, once the package itself has been loaded, by using the `data()` function, and explanatory documentation accessed using the `help()` function, as follows.

```
> data(elevation)
> help(elevation)
```

There are several data-sets included in the package **geoR** which can be loaded with `data()`. Typing the command `data(package="geoR")` will show a list of the available data-sets with respective names and a short description. For each of them there is a help file explaining the data contents and format.

Another, and often more convenient, way of running a sequence of R commands is to use `source()`. To do so, we first type the required sequence of commands, without the `>` at the beginning of each line, into a text file, say `elevation.R`, although any other legal file name could be used. We then invoke the whole sequence by responding to the R prompt with the single command

```
> source("elevation.R")
```

This option, or an equivalent mode of operation based on toggling between an editor and an R command window, is usually more efficient than typing R commands directly in response to the > prompt.

The next example shows the output generated by applying the summary() function to the elevation data. The output includes the number of data points, the minimum and maximum values of the x and y coordinates and of the distances between pairs of points, together with summary statistics for the measured elevations.

```
> summary(elevation)

Number of data points: 52

Coordinates summary
      x   y
min 0.2 0.0
max 6.3 6.2

Distance summary
      min        max
0.200000 8.275869

Data summary
   Min. 1st Qu.  Median    Mean 3rd Qu.    Max.
  690.0   787.5   830.0   827.1   873.0   960.0
```

Another function which is useful for initial exploration of a set of data is the method plot.geodata(), which is invoked by default when a geodata object is supplied as an argument to the built-in plot() function. Its effect is to produce a 2 by 2 display showing the point locations, the measured values at each location against each of the coordinates, and a histogram of the measured values. This plot for the elevation data is shown in Figure 1.8, which is produced by the command

```
> plot(elevation, lowess = T)
```

The optional argument lowess = T adds a smooth curve to the scatterplots of the measured values against each of the spatial coordinates. The top-right panel of Figure 1.8 has been rotated by 90 degrees from the conventional orientation i.e., the measured values correspond to the horizontal rather than the vertical axis so that the spatial coordinate axes have the same interpretation throughout. These plots aim to investigate the behaviour of the data along the coordinates, which can be helpful in deciding whether a trend surface should be included in the model for the data. By default, the plot of the data locations shown in the top-left panel of Figure 1.8 uses circles, triangles, and vertical and diagonal crosses to correspond to the quartiles of the empirical distribution of measured values. On a computer screen, these points would also appear in dif-

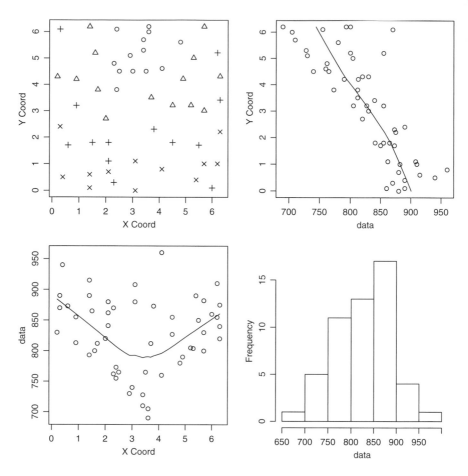

Figure 1.8. Point locations (top left), data values against coordinates (top right and bottom left) and histogram (bottom right) of the measured elevations.

ferent colours: blue, green, yellow and red, respectively. The use of four distinct colours is the default for this function.

1.6.2   More on the *geodata* object

The functions **read.geodata()** and **as.geodata()** store a geostatistical data-set in a particular format called a **geodata** object. A **geodata** object is a list which has two obligatory components: a matrix with the two-dimensional coordinates (**coords**) of the sampling design and a vector giving the corresponding measured value at each of the locations in the design (**data**). Four additional, optional components are: a matrix with coordinates defining the boundary of the polygonal study area (**borders**); a vector or data-frame with covariates (**covariate**); an offset variable (**units.m**); and a vector indexing the number of the realisation of the process if more than one is available (**realisation**), as for instance for data collected at different time points. These additional components, if present, are then used automatically by some of the **geoR** functions.

The example below shows the components of some of the data-sets which are included in the **geoR** package as geodata objects.

```
> names(elevation)

$coords
[1] "x" "y"

$data
[1] "data"

> data(parana)
> names(parana)

$coords
[1] "east"   "north"

$data
[1] "data"

$other
[1] "borders"    "loci.paper"

> data(ca20)
> names(ca20)

$coords
[1] "east"   "north"

$data
[1] "data"

$covariate
[1] "altitude" "area"

$other
[1] "borders" "reg1"    "reg2"    "reg3"

> names(unclass(ca20))

[1] "coords"    "data"         "covariate" "borders"    "reg1"
[6] "reg2"      "reg3"
```

The slightly different results returned from the calls names(ca20) and names(unclass(ca20)) illustrate that some special *methods* have been provided to modify the way that standard R functions handle geodata objects; in this case the standard command names(ca20) recognises that ca20 is a geodata object, and invokes the non-standard method names.geodata(), whereas the command unclass(ca20) gives the standard result of the names function by removing the class geodata from the object ca20.

Other, perhaps more useful methods to facilitate data manipulation are also implemented such as as.data.frame.geodata() which converts a geodata object to a data-frame and subset.geodata() which facilitates extracting subsets of geodata objects. Below we illustrate the usage of subset.geodata() on the ca20 data-set selecting data only within sub-area 3 in the first command and selecting only data greater than 70 in the second.

```
> ca20.3 <- subset(ca20, area == 3)
> ca20.g70 <- subset(ca20, data > 70)
```

1.6.3 Rongelap data

Our next example produces a circle plot for the Rongelap data, together with an enlarged inset of the western part of the island. The rongelap data-set is included with the **geoRglm** package.

```
> require(geoRglm)
> data(rongelap)
```

The response to the command *names(rongelap)* reveals that the Rongelap geodata object has four components: coords contains the spatial coordinates; data contains the photon emission counts y_i attributable to radioactive caesium; units.m is an offset variable which gives the values of t_i, the time (in seconds) over which y_i was accumulated; borders contains the coordinates of a digitisation of the island's coastline. The function summary() recognises and summarises all four components.

```
> names(rongelap)

$coords
NULL

$data
[1] "data"

$units.m
[1] "units.m"

$other
[1] "borders"

> summary(rongelap)

Number of data points: 157

Coordinates summary
      Coord.X Coord.Y
min    -6050   -3430
max     -50       0
```

```
Distance summary
      min        max
   40.000 6701.895
```

```
Borders summary
              [,1]          [,2]
min -6299.31201 -3582.2500
max    20.37916   103.5414
```

```
Data summary
   Min. 1st Qu.  Median   Mean 3rd Qu.    Max.
     75    1975    2639   3011    3437   21390
```

```
Offset variable summary
   Min. 1st Qu.  Median   Mean 3rd Qu.    Max.
  200.0   300.0   300.0  401.9   400.0  1800.0
```

We can use **points()** to visualise the data on a map of the study area as shown in Figure 1.2. For the enlargement of the western part of the island, we have used **subarea()** to select a subset of the original data-set whose spatial coordinates lie within a specified sub-area. The function **subarea()** accepts arguments **xlim** and/or **ylim** defining a rectangular sub-area. If these arguments are not provided the user is prompted to click on two points which then define the opposite corners of the required rectangular area. To produce the figure, we use the following sequence of commands.

```
> points(rongelap)
> rongwest <- subarea(rongelap, xlim = c(-6300, -4800))
> rongwest.z <- zoom.coords(rongwest, xzoom = 3.5, xoff = 2000,
+      yoff = 3000)
> points(rongwest.z, add = T)
> rect.coords(rongwest$sub, lty = 2, quiet = T)
> rect.coords(rongwest.z$sub, lty = 2, quiet = T)
> text(-4000, 1100, "western area", cex = 1.5)
```

The object **rongwest** is a **geodata** object which is generated by **subarea()**. It has the same components as the original **geodata** object but is restricted to the area whose x-coordinates are in the range -6300 to -4800; because the **ylim** argument was not used, the y-coordinate range is unrestricted.

Note that, by default, if the element **units.m** is present in the data object, as for this case, the size of the circle plotted at each location is determined by the corresponding emission count per unit time, rather than by the emission count itself. Setting **data=rongelap$data** the effect of the argument is that the raw data on emission count would be plotted. If preferred, the argument **pt.div="equal"** could be used to specify that all the points should have the same size. The coastline is included in the plot by default because the element **borders** is present in the **geodata** object. If this is unwanted the argument

borders can be set to NULL. Alternatively, another object with the polygon defining the region bondaries can be passed using this argument.

1.6.4   The Gambia malaria data

The Gambia malaria data shown in Example 1.3 are available as a **data-frame** in the **geoR** package. The commands below load the data and display the first three lines of the resulting data-frame, with variable names printed at the head of each column of data.

```
> data(gambia)
> gambia[1:3, ]
```

```
              x       y pos  age netuse treated green phc
1850 349631.3 1458055   1 1783      0       0 40.85   1
1851 349631.3 1458055   0  404      1       0 40.85   1
1852 349631.3 1458055   0  452      1       0 40.85   1
```

Each line corresponds to one child. The columns are the coordinates of the village where the child lives (x and y), whether or not the child tested positive for malaria (*pos*), their age in days (*age*), usage of bed-net(*netuse*), whether the bed-net is treated with insecticide (*treated*), the vegetation index measured at the village location (*green*) and the presence or absence of a health centre in the village (*phc*).

   To display the data as show in Figure 1.3 we use the **gambia.map()** function which is also included in **geoR**.

```
> gambia.map()
```

1.6.5   The soil data

The soil data shown in Example 1.4 are included in **geoR** and can be loaded with the commands **data(ca20)** and **data(camg)**. The former loads only the calcium data, stored as a **geodata** object, whereas the latter loads a data-frame which includes both the calcium and the magnesium data. In order to produce the right-hand panel in Figure 1.5 we use the sequence of commands below.

```
> data(camg)
> mg20 <- as.geodata(camg, data.col = 6)
> points(mg20, cex.min = 0.2, cex.max = 1.5, pch = 21)
> data(ca20)
> polygon(ca20$reg1, lty = 2)
> polygon(ca20$reg2, lty = 2)
> polygon(ca20$reg3, lty = 2)
```

The first command loads the combined data using **data()**, the second creates a **geodata** object for plotting the magnesium data. Borders of the region and sub-regions included in the plot use extra information provided in the calcium data object ca20, which is included in the **geoR** package.

We now inspect the `ca20` object in more detail using the `summary()` function. Remember that *help(ca20)* gives the documentation for this data-set.

```
> summary(ca20)


Number of data points: 178

Coordinates summary
     east north
min 4957  4829
max 5961  5720

Distance summary
       min          max
  43.01163 1138.11774

Borders summary
     east north
min 4920  4800
max 5990  5800

Data summary
   Min. 1st Qu.  Median   Mean 3rd Qu.    Max.
  21.00   43.00   50.50  50.68   58.00   78.00

Covariates summary
      altitude       area
 Min.    :3.300   1: 14
 1st Qu.:5.200    2: 48
 Median :5.650    3:116
 Mean    :5.524
 3rd Qu.:6.000
 Max.    :6.600


Other elements in the geodata object
[1] "reg1" "reg2" "reg3"
```

The output above shows that the data contain 178 locations, with E-W coordinates ranging from 4957 to 5961 and N-S coordinates ranging from 4829 to 5720. The minimum distance between any two locations is about 43 units and the maximum 1138. The object also has a **borders** component which is a two-column matrix with rows corresponding to a set of coordinates defining the polygonal boundary of the study area. The function also shows summary statistics for the response variable and for the covariates. For the covariate **area** the summary indicates that 14, 48 and 116 locations lie within the sub-areas 1, 2 and 3, respectively.

1.7   Exercises

1.1. Produce a plot of the Rongelap data in which a continuous colour scale or grey scale is used to indicate the value of the emission count per unit time at each location, and the two sub-areas with the 5 by 5 sub-grids at 50 metre spacing are shown as insets.

1.2. Construct a polygonal approximation to the boundary of The Gambia. Construct plots of the malaria data which show the spatial variation in the values of the observed prevalence in each village and of the greenness covariate.

1.3. Consider the elevation data as a simple regression problem with elevation as the response and north-south location as the explanatory variable. Fit the standard linear regression model using ordinary least squares. Examine the residuals from the linear model, with a view to deciding whether any more sophisticated treatment of the spatial variation in elevation might be necessary.

1.4. Find a geostatistical data-set which interests you.

   (a) What scientific questions are the data intended to address? Do these concern estimation, prediction, or testing?
   (b) Identify the *study region*, the *design*, the *response* and the *covariates*, if any.
   (c) What is the *support* of each response?
   (d) What is the underlying *signal*?
   (e) If you wished to predict the signal throughout the study region, would you choose to interpolate the response data?

1.5. Load the Paraná data-set using the command **data(parana)** and inspect its documentation using **help(parana)**. For these data, consider the same questions as were raised in Exercise 1.4.

2
An overview of model-based geostatistics

The aim of this chapter is to provide a short overview of model-based geostatistics, using the elevation data of Example 1.1 to motivate the various stages in the analysis. Although this example is very limited from a scientific point of view, its simplicity makes it well suited to the task in hand. Note, however, that Handcock and Stein (1993) show how to construct a useful explanatory variable for these data using a map of streams which run through the study region.

2.1   Design

Statistical design is concerned with deciding what data to collect in order to address a question, or questions, of scientific interest. In this chapter, we shall assume that the scientific objective is to produce a map of surface elevation within a square study region whose side length is 6.7 units, or 335 feet ($\approx$ 102 meters); we presume that this study region has been chosen for good reason, either because it is of interest in its own right, or because it is representative of some wider spatial region.

In this simple setting, there are essentially only two design questions: at how many locations should we measure the elevation and where should we place these locations within the study region?

In practice, the answer to the first question is usually dictated by limits on the investigator's time or any additional cost in converting each field sample into a measured value. For example, some kinds of measurements involve expensive off-site laboratory assays, whereas others, such as surface elevation, can be measured directly in the field. For whatever reason, the answer in this example is 52.

For the second question, two obvious candidate designs are a *completely random* design or a *completely regular* design. In the former, the locations x_i form an independent random sample from the uniform distribution over the study area, that is a homogeneous planar Poisson process (Diggle 2003, chapter 1). In the latter, the x_i form a regular lattice pattern over the study region. Classical sampling theory (Cochran, 1977) tends to emphasise the virtue of some form of random sampling to ensure unbiased estimation of underlying population characteristics, whereas spatial sampling theory (Matérn, 1960) shows that under typical modelling assumptions spatial properties are more efficiently estimated by a regular design. A compromise, which the originators of the surface elevation data appear to have adopted, is to use a design which is more regular than the completely random design but not as regular as a lattice.

Lattice designs are widely used in applications. The convenience of lattice designs for fieldwork is obvious, and provided there is no danger that the spacing of the lattice will match an underlying periodicity in the spatial phenomenon being studied, lattice designs are generally efficient for spatial prediction (Matérn, 1960). In practice, the rigidity and simplicity of a lattice design also provide some protection against sub-conscious bias in the placing of the x_i. Note in this context that, strictly, a regular lattice design should mean a lattice whose origin is located at random, to guard against any subjective bias. The soil data of Example 1.4 provide an example of a regular lattice design.

Even more common in some areas of application is the *opportunistic design*, whereby geostatistical data are collected and analysed using an existing network of locations x_i which may have been established for quite different purposes. Designs of this kind often arise in connection with environmental monitoring. In this context, individual recording stations may be set up to monitor pollution levels from particular industrial sources or in environmentally sensitive locations, without any thought initially that the resulting data might be combined in a single, spatial analysis. This immediately raises the possibility that the design may be preferential, in the sense discussed in Section 1.2.3. Whether they arise by intent or by accident, preferential designs run the risk that a standard geostatistical analysis may produce misleading inferences about the underlying continuous spatial variation.

2.2   Model formulation

We now consider model formulation — unusually before, rather than after, exploratory data analysis. In practice, clean separation of these two stages is rare. However, in our experience it is useful to give some consideration to the kind of model which, in principle, will address the questions of interest before refining the model through the usual iterative process of data analysis followed by reformulation of the model as appropriate.

For the surface elevation data, the scientific question is a simple one — how can we use the measured elevations to construct our best guess (or, in more formal language, to predict) the underlying elevation surface throughout the study region? Hence, our model needs to include a real-valued, spatially continuous stochastic process, $S(x)$ say, to represent the surface elevation as a function of location, x. Depending on the nature of the terrain, we may want $S(x)$ to be continuous, differentiable or many-times differentiable. Depending on the nature of the measuring device, or the skill of its operator, we may also want to allow for some discrepancy between the true surface elevation $S(x_i)$ and the measured value Y_i at the design location x_i. The simplest statistical model which meets these requirements is a stationary Gaussian model, which we define below. Later, we will discuss some of the many possible extensions of this model which increase its flexibility.

We denote a set of geostatistical data in its simplest form i.e., in the absence of any explanatory variables, by $(x_i, y_i) : i = 1, \ldots, n$ where the x_i are spatial locations and y_i is the measured value associated with the location x_i. The assumptions underlying the stationary Gaussian model are:

1. $\{S(x) : x \in \mathbb{R}^2\}$ is a Gaussian process with mean μ, variance $\sigma^2 = \mathrm{Var}\{S(x)\}$ and correlation function $\rho(u) = \mathrm{Corr}\{S(x), S(x')\}$, where $u = ||x - x'||$ and $|| \cdot ||$ denotes distance;

2. conditional on $\{S(x) : x \in \mathbb{R}^2\}$, the y_i are realisations of mutually independent random variables Y_i, normally distributed with conditional means $\mathrm{E}[Y_i|S(\cdot)] = S(x_i)$ and conditional variances τ^2.

The model can be defined equivalently as

$$Y_i = S(x_i) + Z_i : i = 1, \ldots, n$$

where $\{S(x) : x \in \mathbb{R}^2\}$ is defined by assumption 1 above and the Z_i are mutually independent $N(0, \tau^2)$ random variables. We favour the superficially more complicated conditional formulation for the joint distribution of the Y_i given the signal, because it identifies the model explicitly as a special case of the generalized linear geostatistical model which we introduced in Section 1.4.

In order to define a legitimate model, the correlation function $\rho(u)$ must be positive-definite. This condition imposes non-obvious constraints so as to ensure that, for any integer m, set of locations x_i and real constants a_i, the linear combination $\sum_{i=1}^{m} a_i S(x_i)$ will have non-negative variance. In practice, this is usually ensured by working within one of several standard classes of parametric model for $\rho(u)$. We return to this question in Chapter 3. For the moment, we note only that a flexible, two-parameter class of correlation functions due to Matérn (1960) takes the form

$$\rho(u; \phi, \kappa) = \{2^{\kappa-1}\Gamma(\kappa)\}^{-1}(u/\phi)^{\kappa} K_{\kappa}(u/\phi) \tag{2.1}$$

where $K_{\kappa}(\cdot)$ denotes the modified Bessel function of the second kind, of order κ. The parameter $\phi > 0$ determines the rate at which the correlation decays to zero with increasing u. The parameter $\kappa > 0$ is called the *order* of the Matérn model, and determines the differentiability of the stochastic process $S(x)$, in a sense which we shall make precise in Chapter 3.

Our notation for $\rho(u)$ presumes that $u \geq 0$. However, the correlation function of any stationary process must by symmetric in u, hence $\rho(-u) = \rho(u)$.

The stochastic variation in a physical quantity is not always well described by a Gaussian distribution. One of the simplest ways to extend the Gaussian model is to assume that the model holds after applying a transformation to the original data. For positive-valued response variables, a useful class of transformations is the Box-Cox family (Box and Cox, 1964):

$$Y^* = \begin{cases} (Y^\lambda - 1)/\lambda & : \quad \lambda \neq 0 \\ \log Y & : \quad \lambda = 0 \end{cases} \tag{2.2}$$

Another simple extension to the basic model is to allow a spatially varying mean, for example by replacing the constant μ by a linear regression model for the conditional expectation of Y_i given $S(x_i)$, so defining a spatially varying mean $\mu(x)$.

A third possibility is to allow $S(x)$ to have non-stationary covariance structure. Arguably, most spatial phenomena exhibit some form of non-stationarity, and the stationary Gaussian model should be seen only as a convenient approximation to be judged on its usefulness rather than on its strict scientific provenance.

2.3 Exploratory data analysis

Exploratory data analysis is an integral part of modern statistical practice, and geostatistics is no exception. In the geostatistical setting, exploratory analysis is naturally oriented towards the preliminary investigation of spatial aspects of the data which are relevant to checking whether the assumptions made by any provisional model are approximately satisfied. However, non-spatial aspects can and should also be investigated.

2.3.1 Non-spatial exploratory analysis

For the elevation data in Example 1.1 the 52 data values range from 690 to 960, with mean 827.1, median 830 and standard deviation 62. A histogram of the 52 elevation values (Figure 2.1) indicates only mild asymmetry, and does not suggest any obvious outliers. This adds some support to the use of a Gaussian model as an approximation for these data. Also, because geostatistical data are, at best, a correlated sample from a common underlying distribution, the shape of their histogram will be less stable than that of an independent random sample of the same size, and this limits the value of the histogram as a diagnostic for non-normality.

In general, an important part of exploratory analysis is to examine the relationship between the response and available covariates, as illustrated for the

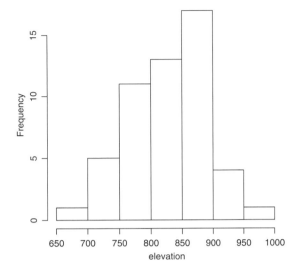

Figure 2.1. Histogram of the surface elevation data.

soil data in Figure 1.7. For the current example, the only available covariates
to consider are the spatial coordinates themselves.

2.3.2 Spatial exploratory analysis

The first stage in spatial exploratory data analysis is simply to plot the response
data in relation to their locations, for example using a circle plot as shown for
the surface elevation data in Figure 1.1. Careful inspection of this plot can
reveal spatial outliers i.e., responses which appear grossly discordant with their
spatial neighbours, or spatial trends which might suggest the need to include
a trend surface model for a spatially varying mean, or perhaps qualitatively
different behaviour in different sub-regions.

In our case, the most obvious feature of Figure 1.1 is the preponderance of
large response values towards the southern end of the study region. This suggests that a trend surface term in the model might be appropriate. In some
applications, the particular context of the data might suggest that there is
something special about the north-south direction — for example, for applications on a large geographical scale, we might expect certain variables relating
to the physical environment to show a dependence on latitude. Otherwise, our
view would be that if a trend surface is to be included in the model at all, then
both of the spatial coordinates should contribute to it because the orientation
of the study region is essentially arbitrary.

Scatterplots of the response variable against each of the spatial coordinates
can sometimes reveal spatial trends more clearly. Figure 2.2 show the surface elevations plotted against each of the coordinates, with lowess smooths (Cleveland,
1979, 1981) added to help visualisation. These plots confirm the north-south
trend whilst additionally suggesting a less pronounced, non-monotone east-west

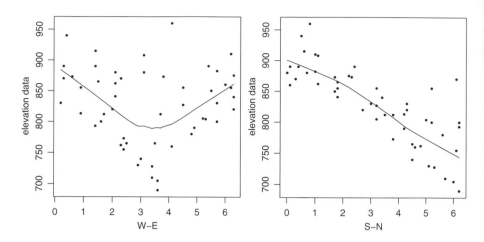

Figure 2.2. Elevation data against the coordinates.

trend, with higher responses concentrated towards the eastern and western edges of the study region.

When interpreting plots of this kind it can be difficult, especially when analysing small data-sets, to distinguish between a spatially varying mean response and correlated spatial variation about a constant mean. Strictly speaking, without independent replication the distinction between a deterministic function $\mu(x)$ and the realisation of a stochastic process $S(x)$ is arbitrary. Operationally, we make the distinction by confining ourselves to "simple" functions $\mu(x)$, for example low-order polynomial trend surfaces, using the correlation structure of $S(x)$ to account for more subtle patterns of spatial variation in the response. In Chapter 5 we shall use formal, likelihood-based methods to guide our choice of model for both mean and covariance structure. Less formally, we interpret spatial effects which vary on a scale comparable to or greater than the dimensions of the study region as variation in $\mu(x)$ and smaller-scale effects as variation in $S(x)$. This is in part a pragmatic strategy, since covariance functions which do not decay essentially to zero at distances shorter than the dimensions of the study region will be poorly identified, and in practice indistinguishable from spatial trends. Ideally, the model for the trend should also have a natural physical interpretation; for example, in an investigation of the dispersal of pollutants around a known source, it would be natural to model $\mu(x)$ as a function of the distance, and possibly the orientation, of x relative to the source.

To emphasise this point, the three panels of Figure 2.3 compare the original Figure 1.1 with circle plots of residuals after fitting linear and quadratic trend surface models by ordinary least squares. If we assume a constant spatial mean for the surface elevations themselves, then the left-hand panel of Figure 2.3 indicates that the elevations must be very strongly spatially correlated, to the extent that the correlation persists at distances beyond the scale of the study region. As noted above, fitting a model of this kind to the data would result in poor identification of parameters describing the correlation structure. If, in

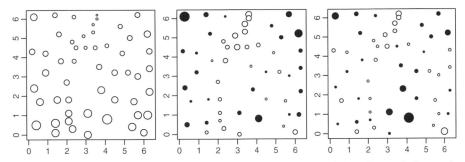

Figure 2.3. Circle plot of the surface elevation data. The left-hand panel shows the original data. The centre and right-hand panels show the residuals from first-order (linear) and second-order (quadratic) polynomial trend surfaces, respectively, using empty and filled circles to represent negative and positive residuals and circle radii proportional to the absolute values of the residuals.

contrast, we use a linear trend surface to describe a spatially varying mean, then the central panel of Figure 2.3 still suggests spatial correlation because positive and negative residuals tend to occur together, but the scale of the spatial correlation is smaller. The right-hand panel of 2.3 has a qualitatively similar appearance to the centre panel, but the range of the residuals has been reduced, because some additional variation is taken up by the quadratic terms in the fitted trend surface. The range of the residuals is from -61.1 to $+110.7$ in the centre panel, and from -63.3 to $+97.8$ in the right-hand panel.

Notwithstanding the above discussion, visual assessment of spatial correlation from a circle plot is difficult. For a sharper assessment, a useful exploratory tool is the *empirical variogram*. We discuss theoretical and empirical variograms in more detail in Chapters 3 and 5, respectively. Here, we give only a brief description.

For a set of geostatistical data $(x_i, y_i) : i = 1, \ldots, n$, the *empirical variogram ordinates* are the quantities $v_{ij} = \frac{1}{2}(y_i - y_j)^2$. For obvious reasons, some authors refer to these as the *semi-variogram ordinates*. If the y_i have spatially constant mean and variance, then v_{ij} has expectation $\sigma^2\{1 - \rho(x_i, x_j)\}$ where σ^2 is the variance and $\rho(x_i, x_j)$ denotes the correlation between y_i and y_j. If the y_i are generated by a stationary spatial process, then $\rho(\cdot)$ depends only on the distance between x_i and x_j and typically approaches zero at large distances, hence the expectation of the v_{ij} approaches a constant value, σ^2, as the distance u_{ij} between x_i and x_j increases. If the y_i are uncorrelated, then all of the v_{ij} have expectation σ^2. These properties motivate the definition of the *empirical variogram* as a plot of v_{ij} against the corresponding distance u_{ij}. A more easily interpretable plot is obtained by averaging the v_{ij} within distance bands.

The left-hand panel of Figure 2.4 shows a variogram for the original surface elevations, whilst the right-hand panel shows variograms for residuals from the linear and quadratic trend surface models, indicated by solid and dashed lines, respectively. In the left-hand panel, the variogram increases throughout the

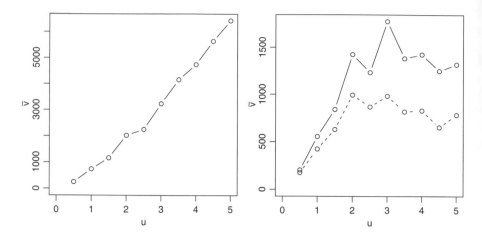

Figure 2.4. Empirical variograms for the original data (left panel) and for residuals (right panel) from a linear (solid lines) or quadratic (dashed lines) trend surface. In all three cases, empirical variogram ordinates have been averaged in bins of unit width.

plotted range, indicating that *if* these data were generated by a stationary stochastic process, then the range of its spatial correlation must extend beyond the scale of the study region. Pragmatically, including a spatially varying mean is a better modelling strategy. The solid line on the right-hand panel shows behaviour more typical of a stationary, spatially correlated process i.e., an initial increase levelling off as the correlation decays to zero at larger distances. Finally, the shape of the variogram in the dashed line on the right-hand panel is similar to the solid one but its range is smaller by a factor of about 0.6. The range of values in the ordinates of the empirical variogram is approximately equal to the variance of the residuals, hence the reduction in range again indicates how the introduction of progressively more elaborate models for the mean accounts for correspondingly more of the empirical variation in the original data. Note also that in both panels of Figure 2.4 the empirical variogram approaches zero at small distances. This indicates that surface elevation is being measured with negligible error, relative to either the spatial variation in the surface elevation itself (left-hand panel), or the residual spatial variation about the linear or quadratic trend surface (right-hand panel). This interpretation follows because the expectation of v_{ij} corresponding to two independent measurements, y_i and y_j, at the same location is simply the variance of the measurement error.

We emphasise that, for reasons explained in Chapter 5, we prefer to use the empirical variogram only as an exploratory tool, rather than as the basis for formal inference. With this proviso, Figure 2.4 gives a strong indication that a stationary model is unsuitable for these data, whereas the choice between the linear and quadratic trend surface models is less clear-cut.

When an empirical variogram appears to show little or no spatial correlation, it can be useful to assess more formally whether the data are compatible with an underlying model of the form $y_i = \mu(x_i) + z_i$ where the z_i are uncorrelated residuals about a spatially varying mean $\mu(x)$. A simple way to do

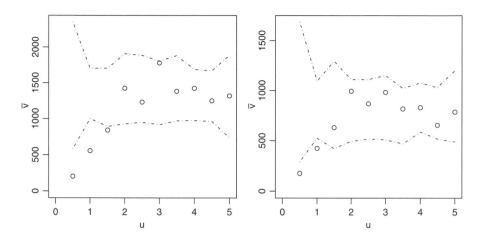

Figure 2.5. Monte Carlo envelopes for the variogram of ordinary least squares residuals of the surface elevation data after fitting linear (left-hand panel) or quadratic
(right-hand panel) trend surface models.

this is to compute residuals about a fitted mean $\hat{\mu}(x)$ and to compare the
residual empirical variogram with the envelope of empirical variograms computed from random permutations of the residuals, holding the corresponding
locations fixed. The left-hand panel of Figure 2.5 shows a variogram envelope
obtained from 99 independent random permutations of the residuals from a
linear trend surface fitted to the surface elevations by ordinary least squares.
This shows that the increasing trend in the empirical variogram is statistically
significant, confirming the presence of positive spatial correlation. The same
technique applied to the residuals from the quadratic trend surface produces
the diagram shown as the right-hand panel of Figure 2.5. This again indicates
significant spatial correlation, although the result is less clear-cut than before,
as the empirical variogram ordinates at distances 0.5 and 1.0 fall much closer
to the lower simulation envelope than they do in the left-hand panel.

2.4   The distinction between parameter estimation and spatial prediction

Before continuing with our illustrative analysis of the surface elevation data, we
digress to expand on the distinction between estimation and prediction.

Suppose that $S(x)$ represents the level of air pollution at the location x,
that we have observed (without error, in this hypothetical example) the values
$S_i = S(x_i)$ at a set of locations $x_i : i = 1, \ldots, n$ forming a regular lattice over a
spatial region of interest, A, and that we wish to learn about the average level
of pollution over the region A. An intuitively reasonable estimate is the sample

mean,

$$\bar{S} = n^{-1} \sum_{i=1}^{n} S_i. \tag{2.3}$$

What precision should we attach to this estimate?

Suppose that $S(x)$ has a constant expectation, $\theta = \mathrm{E}[S(x)]$ for any location x in A. One possible interpretation of $\bar{S}$ is as an estimate of θ, in which case an appropriate measure of precision is the mean square error, $\mathrm{E}[(\bar{S} - \theta)^2]$. This is just the variance of $\bar{S}$, which we can calculate as

$$n^{-2} \sum_{i=1}^{n} \sum_{j=1}^{n} \mathrm{Cov}(S_i, S_j). \tag{2.4}$$

For a typical geostatistical model, the correlation between any two S_i and S_j will be either zero or positive, and (2.4) will therefore be larger than the naive expression for the variance of a sample mean, σ^2/n where $\sigma^2 = \mathrm{Var}\{S(x)\}$.

If we regard $\bar{S}$ as a *predictor* of the *spatial average*,

$$S_A = |A|^{-1} \int_A S(x)dx,$$

where $|A|$ is the area of A, then the mean square prediction error is $\mathrm{E}[(\bar{S} - S_A)^2]$. Noting that S_A is a random variable, we write this as

$$
\begin{aligned}
\mathrm{E}[(\bar{S} - S_A)^2] \quad = \quad & n^{-2} \sum_{i=1}^{n} \sum_{j=1}^{n} \mathrm{Cov}(S_i, S_j) \\
+ \quad & |A|^{-2} \int_A \int_A \mathrm{Cov}\{S(x), S(x')\}dxdx' \\
- \quad & 2(n|A|)^{-1} \sum_{i=1}^{n} \int_A \mathrm{Cov}\{S(x), S(x_i)\}dx. \tag{2.5}
\end{aligned}
$$

In particular, the combined effect of the second and third terms on the right-hand side of (2.5) can easily be to make the mean square prediction error smaller than the naive variance formula. For example, if we increase the sample size n by progressively decreasing the spacing of the lattice points x_i, (2.5) approaches zero, whereas (2.4) does not.

2.5   Parameter estimation

For the stationary Gaussian model, the parameters to be estimated are the mean μ and any additional parameters which define the covariance structure of the data. Typically, these include the signal variance σ^2, the conditional or measurement error variance τ^2 and one or more correlation function parameters ϕ.

In geostatistical practice, these parameters can be estimated in a number of different ways which we shall discuss in detail in Chapter 5. Our preference here is to use the method of maximum likelihood within the declared Gaussian model.

For the elevation data, if we assume a stationary Gaussian model with a Matérn correlation function and a fixed value $\kappa = 1.5$, the maximum likelihood estimates of the remaining parameters are $\hat{\mu} = 848.3$, $\hat{\sigma}^2 = 3510.1$, $\hat{\tau}^2 = 48.2$ and $\hat{\phi} = 1.2$.

However, our exploratory analysis suggested a model with a non-constant mean. Here, we assume a linear trend surface,

$$\mu(x) = \beta_0 + \beta_1 d_1 + \beta_2 d_2$$

where d_1 and d_2 are the north-south and east-west coordinates. In this case the parameter estimates are $\hat{\beta}_0 = 912.5$, $\hat{\beta}_1 = -5$, $\hat{\beta}_2 = -16.5$, $\hat{\sigma}^2 = 1693.1$, $\hat{\tau}^2 = 34.9$ and $\hat{\phi} = 0.8$. Note that because the trend surface accounts for some of the spatial variation, the estimate of σ^2 is considerably smaller than for the stationary model, and similarly for the parameter ϕ which corresponds to the range of the spatial correlation. As anticipated, for either model the estimate of τ^2 is much smaller than the estimate of σ^2. The ratio of $\hat{\tau}^2$ to $\hat{\sigma}^2$ is 0.014 for the stationary model, and 0.021 for the linear trend surface model.

2.6 Spatial prediction

For prediction of the underlying, spatially continuous elevation surface we shall here illustrate perhaps the simplest of all geostatistical methods: *simple kriging*. In our terms, simple kriging is minimum mean square error prediction under the stationary Gaussian model, but ignoring parameter uncertainty i.e., estimates of all model parameters are plugged into the prediction equations as if they were the true parameter values. As discussed earlier, we do not claim that this is a good model for the surface elevation data.

The minimum mean square error predictor, $\hat{S}(x)$ say, of $S(x)$ at an arbitrary location x is the function of the data, $y = (y_1, \ldots, y_n)$, which minimises the quantity $E[\{\hat{S}(x) - S(x)\}^2]$. A standard result, which we discuss in Chapter 6, is that $\hat{S}(x) = E[S(x)|y]$. For the stationary Gaussian process, this conditional expectation is a linear function of the y_i, namely

$$\hat{S}(x) = \mu + \sum_{i=1}^{n} w_i(x)(y_i - \mu) \tag{2.6}$$

where the $w_i(x)$ are explicit functions of the covariance parameters σ^2, τ^2 and ϕ.

The top-left panel of Figure 2.6 gives the result of applying (2.6) to the surface elevation data, using as values for the model parameters the maximum likelihood estimates reported in Section 2.5, whilst the bottom-left panel shows the corresponding prediction standard errors, $SE(x) = \sqrt{\text{Var}\{S(x)|y\}}$. The predictions follow the general trend of the observed elevations whilst smoothing out local irregularities. The prediction variances are generally small at locations close to the sampling locations, because $\hat{\tau}^2$ is relatively small; had we used the

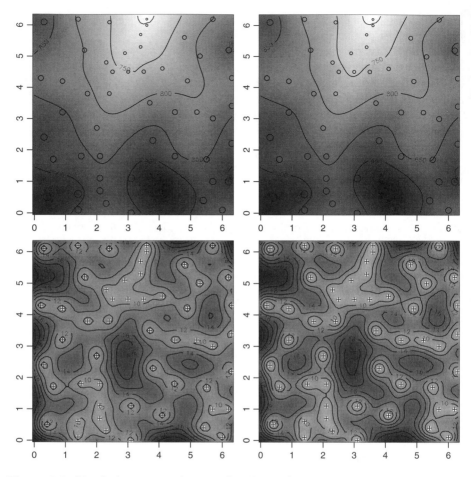

Figure 2.6. Simple kriging predictions for the surface elevation data. The top-left panel shows the simple kriging predictor as a grey-scale image and contour plot; sampling locations are plotted as circles with radii proportional to observed elevations. The bottom-left panel shows the prediction standard deviations; sampling locations are plotted as small crosses. The top-right and bottom-right panels give the same information, but based on the model with a linear trend surface.

value $\tau^2 = 0$ the prediction standard error would have been exactly zero at each sampling location and the predicted surface $\hat{S}(x)$ would have interpolated the observed responses y_i.

It is straightforward to adapt the simple kriging formula (2.6) to incorporate a spatially varying mean. We simply replace the constant μ on the right-hand-side of (2.6) by a spatial trend, $\mu(x)$. If we do this, using the linear trend surface model and its associated maximum likelihood parameter estimates we obtain the results summarised in the top-right and bottom-right panels of Figure 2.6. The plots corresponding to the two different models are directly comparable because they use a common grey scale within each pair. Note in particular that in this simple example, the dubious assumption of stationarity has not prevented the simple kriging methodology from producing a predicted surface which captures qualitatively the apparent spatial trend in the data, and which is almost identical to the predictions obtained using the more reasonable linear trend surface model. The two models produce somewhat different prediction standard errors; these range between 0 and 25.5 for the stationary model, between 0 and 24.4 for the model with the linear trend surface and between 0 and 22.9 for the model with the quadratic trend surface. The differences amongst the three models are rather small. They are influenced by several different aspects of the data and model, including the data configuration and the estimated values of the model parameters. In other applications, the choice of model may have a stronger impact on the predictive inferences we make from the data, even when this choice does not materially affect the point predictions of the underlying surface $S(x)$. Note also that the plug-in standard errors quoted here do not account for parameter uncertainty.

2.7 Definitions of distance

A fundamental stage in any geostatistical analysis is to define the metric for calculating the distance between any two locations. By default, we use the standard planar Euclidean distance i.e., the "straight-line distance" between two locations in $\mathbb{R}^2$. Non-Euclidean metrics may be more appropriate for some applications. For example, Rathbun (1998) discusses the measurement of distance between points in an estuarine environment where, arguably, two locations which are close in the Euclidean metric but separated by dry land should not be considered as near neighbours. It is not difficult to think of other settings where natural barriers to communication might lead the investigator to question whether it is reasonable to model spatial correlation in terms of straight-line distance.

Even when straight-line distance is an appropriate metric, if the study region is geographically extensive, distances computed between points on the earth's surface should strictly be great-circle distances, rather than straight-line distances on a map projection. Using (θ, ϕ) to denote a location in degrees of longitude and latitude, and treating the earth as a sphere of radius $r = 6378$ kilometres, the great-circle distance between two locations is

$$r \cos^{-1}\{\sin\phi_1 \sin\phi_2 + \cos\phi_1 \cos\phi_2 \cos(\theta_1 - \theta_2)\}.$$

section 3.2 of Waller and Gotway (2004) gives a nice discussion of this issue from a statistical perspective. Banerjee (2005) examines the effect of distance computations on geostatistical analysis and concludes that the choice of metric may influence the resulting inferences, both for parameter estimation and for prediction. Note in particular that degrees of latitude and longitude represent approximately equal distances only close to the equator.

Distance calculations are especially relevant to modelling spatial correlation, hence parameters which define the correlation structure are particularly sensitive to the choice of metric. Furthermore, the Euclidean metric plays an integral part in determining valid classes of correlation functions using Bochner's theorem (Stein, 1999). Our **geoR** software implementation only calculates planar Euclidean distances.

2.8 Computation

The non-spatial exploratory analysis of the surface elevation data reported in this chapter uses only built-in R functions as follows.

```
> with(elevation, hist(data, main = "", xlab = "elevation"))
> with(elevation, plot(coords[, 1], data, xlab = "W-E",
+       ylab = "elevation data", pch = 20, cex = 0.7))
> lines(lowess(elevation$data ~ elevation$coords[, 1]))
> with(elevation, plot(coords[, 2], data, xlab = "S-N",
+       ylab = "elevation data", pch = 20, cex = 0.7))
> lines(with(elevation, lowess(data ~ coords[, 2])))
```

To produce circle plots of the residual data we use the **geoR** function `points.geodata()`, which is invoked automatically when a `geodata` object is passed as an argument to the built-in function `points()`, as indicated below. The argument `trend` defines a linear model on the covariates from which the residuals are extracted for plotting. The values `"1st"` and `"2nd"` passed to the argument `trend` are aliases to indicate first-degree and second-degree polynomials on the coordinates. More details and other options to specify the trend are discussed later in this section and in the documentation for `trend.spatial()`. Setting `abs=T` instructs the function to draw the circles with radii proportional to the absolute values of the residuals.

```
> points(elevation, cex.max = 2.5)
> points(elevation, trend = "1st", pt.div = 2, abs = T,
+       cex.max = 2.5)
> points(elevation, trend = "2nd", pt.div = 2, abs = T,
+       cex.max = 2.5)
```

To calculate and plot the empirical variograms shown in Figure 2.4 for the original data and for the residuals, we use `variog()`. The argument `uvec` defines the classes of distance used when computing the empirical variogram, whilst `plot()` recognises that its argument is a variogram object, and automatically invokes `plot.variogram()`. The argument `trend` is used to indicate that the variogram should be calculated from the residuals about a fitted trend surface.

```
> plot(variog(elevation, uvec = seq(0, 5, by = 0.5)),
+       type = "b")
> res1.v <- variog(elevation, trend = "1st", uvec = seq(0,
+       5, by = 0.5))
> plot(res1.v, type = "b")
> res2.v <- variog(elevation, trend = "2nd", uvec = seq(0,
+       5, by = 0.5))
> lines(res2.v, type = "b", lty = 2)
```

To obtain the residual variogram and simulation envelopes under random permutation of the residuals, as shown in Figure 2.5, we proceed as in the following example. By default, the function uses 99 simulations, but this can be changed using the optional argument nsim.

```
> set.seed(231)
> mc1 <- variog.mc.env(elevation, obj = res1.v)
> plot(res1.v, env = mc1, xlab = "u")
> mc2 <- variog.mc.env(elevation, obj = res2.v)
> plot(res2.v, env = mc2, xlab = "u")
```

To obtain maximum likelihood estimates of the Gaussian model, with or without a trend term, we use the **geoR** function likfit(). Because this function uses a numerical maximisation procedure, the user needs to provide initial values for the covariance parameters, using the argument ini. In this example we use the default value 0 for the parameter τ^2, in which case ini specifies initial values for the parameters σ^2 and ϕ. Initial values are not required for the mean parameters.

```
> ml0 <- likfit(elevation, ini = c(3000, 2), cov.model = "matern",
+     kappa = 1.5)
> ml0

likfit: estimated model parameters:
       beta        tausq      sigmasq          phi
" 848.317" "   48.157" "3510.096" "    1.198"

likfit: maximised log-likelihood = -242.1

> ml1 <- likfit(elevation, trend = "1st", ini = c(1300,
+     2), cov.model = "matern", kappa = 1.5)
> ml1

likfit: estimated model parameters:
       beta0        beta1        beta2        tausq      sigmasq
" 912.4865" "   -4.9904" "  -16.4640" "   34.8953" "1693.1329"
         phi
"    0.8061"

likfit: maximised log-likelihood = -240.1
```

To carry out the spatial interpolation using simple kriging we first define, and store in the object locs, a grid of locations at which predictions of the values of the underlying surface are required. The function krige.control() then defines the model to be used for the interpolation, which is carried out by krige.conv(). In the example below, we first obtain predictions for the stationary model, and then for the model with a linear trend on the coordinates. If required, the user can restrict the trend surface model, for example by specifying a linear trend is the north-south direction. However, as a general rule we prefer our inferences to be invariant to the particular choice of coordinate axes, and would therefore fit both linear trend parameters or, more generally, full polynomial trend surfaces.

```
> locs <- pred_grid(c(0, 6.3), c(0, 6.3), by = 0.1)
> KC <- krige.control(type = "sk", obj.mod = ml0)
> sk <- krige.conv(elevation, krige = KC, loc = locs)
> KCt <- krige.control(type = "sk", obj.mod = ml1, trend.d = "1st",
+      trend.l = "1st")
> skt <- krige.conv(elevation, krige = KCt, loc = locs)
```

Finally, we use a selection of built-in graphical functions to produce the maps shown in Figure 2.6, using optional arguments to the graphical functions to ensure that pairs of corresponding plots use the same grey scale.

```
> pred.lim <- range(c(sk$pred, skt$pred))
> sd.lim <- range(sqrt(c(sk$kr, skt$kr)))
> image(sk, col = gray(seq(1, 0, l = 51)), zlim = pred.lim)
> contour(sk, add = T, nlev = 6)
> points(elevation, add = TRUE, cex.max = 2)
> image(skt, col = gray(seq(1, 0, l = 51)), zlim = pred.lim)
> contour(skt, add = T, nlev = 6)
> points(elevation, add = TRUE, cex.max = 2)
> image(sk, value = sqrt(sk$krige.var), col = gray(seq(1,
+      0, l = 51)), zlim = sd.lim)
> contour(sk, value = sqrt(sk$krige.var), levels = seq(10,
+      27, by = 2), add = T)
> points(elevation$coords, pch = "+")
> image(skt, value = sqrt(skt$krige.var), col = gray(seq(1,
+      0, l = 51)), zlim = sd.lim)
> contour(skt, value = sqrt(skt$krige.var), levels = seq(10,
+      27, by = 2), add = T)
> points(elevation$coords, pch = "+")
```

In **geoR**, covariates which define a linear model for the mean response can be specified by passing additional arguments to plotting or model-fitting functions. In the examples above, we used trend="1st" or trend="2nd" to specify a linear or quadratic trend surface. However, these are simply short-hand aliases to formulae which define the corresponding linear models, and are provided for users' convenience. For example, the *model formula* trend=~coords[,1] + coords[,2] would produce the same result as trend="1st". The trend argument will also accept a matrix representing the design matrix of a general linear model, or the output of the trend definition function, trend.spatial(). For example, the call below to plot() can be used in order to inspect the data after taking out the linear effect of the north-south coordinate. By setting the argument trend=~coords[,2] the function fits a standard linear model on this covariate and uses the residuals to produce the plots shown in Figure 2.7, rather than plotting the original response data. Similarly, we could fit a quadratic function on the x-coordinate by setting trend=~coords[,2] + poly(coords[,1], degree=2). We invite the reader to experiment with different options for the

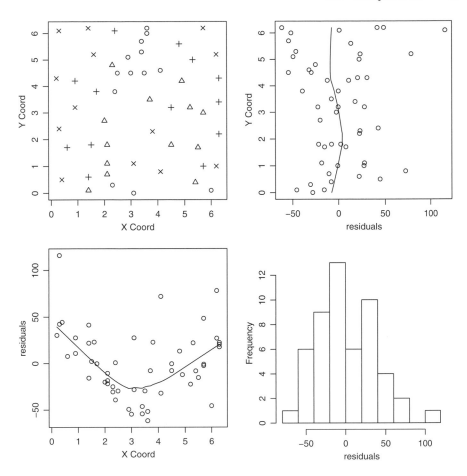

Figure 2.7. Output of plot.geodata() when setting the argument trend=~coords[,2].

argument trend and trend.spatial(). The procedure of taking out the effect of a covariate is sometimes called *trend removal*.

```
> plot(elevation, low = TRUE, trend = ~coords[, 2], qt.col = 1)
```

The trend argument can also be used to take account of covariates other than functions of the coordinates. For example, the data-set ca20 included in **geoR** stores the calcium content from soil samples, as discussed in Example 1.4, together with associated covariate information. Recall that in this example the study region is divided in three sub-regions with different histories of soil management. The covariate area included in the data-set indicates for each datum the sub-region in which it was collected. Figure 2.8 shows the exploratory plot for the residuals after removing a separate mean for calcium content in each sub-region. This diagram was produced using the following code.

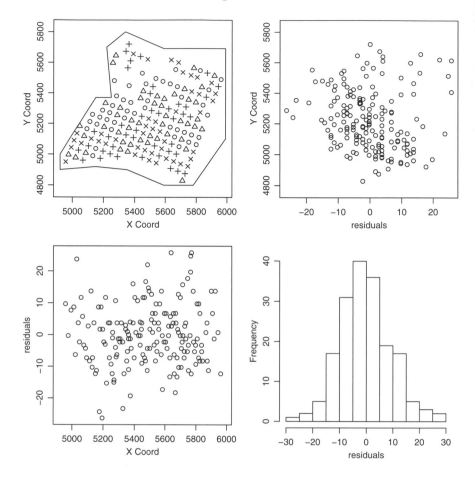

Figure 2.8. Exploratory plot for the `ca20` data-set obtained when setting `trend=~area`.

```
> data(ca20)
> plot(ca20, trend = ~area, qt.col = 1)
```

The plotting functions in `geoR` also accept an optional argument `lambda` which specifies the numerical value for the parameter of the Box-Cox family of transformations, with default `lambda=1` corresponding to no transformation. For example, the command

```
> plot(ca20, lambda = 0)
```

sets the Box-Cox transformation parameter to $\lambda = 0$, which will then produce plots using the logarithm of the original response variable.

2.9   Exercises

2.1. Investigate the R packages **splancs** or **spatstat**, both of which provide functions for the analysis of spatial point pattern data. Use either of these packages to confirm (or not, as the case may be) that the design used for the surface elevation data is more regular than a completely random design.

2.2. Consider the following two models for a set of responses, $Y_i : i = 1, \ldots, n$ associated with a sequence of positions $x_i : i = 1, \ldots, n$ along a one-dimensional spatial axis x.

(a) $Y_i = \alpha + \beta x_i + Z_i$, where α and β are parameters and the Z_i are mutually independent with mean zero and variance σ_Z^2.

(b) $Y_i = A + B x_i + Z_i$ where the Z_i are as in (a) but A and B are now random variables, independent of each other and of the Z_i, each with mean zero and respective variances σ_A^2 and σ_B^2.

For each of these models, find the mean and variance of Y_i and the covariance between Y_i and Y_j for any $j \neq i$. Given a single realisation of either model, would it be possible to distinguish between them?

2.3. Suppose that $Y = (Y_1, \ldots, Y_n)$ follows a multivariate Gaussian distribution with $E[Y_i] = \mu$ and $\text{Var}\{Y_i\} = \sigma^2$ and that the covariance matrix of Y can be expressed as $V = \sigma^2 R(\phi)$. Write down the log-likelihood function for $\theta = (\mu, \sigma^2, \phi)$ based on a single realisation of Y and obtain explicit expressions for the maximum likelihood estimators of μ and σ^2 when ϕ is known. Discuss how you would use these expressions to find maximum likelihood estimators numerically when ϕ is unknown.

2.4. Load the `ca20` data-set with `data(ca20)`. Check the data-set documentation with `help(ca20)`. Perform an exploratory analysis of these data. Would you include a trend term in the model? Would you recommend a data transformation? Is there evidence of spatial correlation?

2.5. Load the Paraná data with `data(parana)` and repeat Exercise 2.4.

3
Gaussian models for geostatistical data

Gaussian stochastic processes are widely used in practice as models for geostatistical data. These models rarely have any physical justification. Rather, they are used as convenient empirical models which can capture a wide range of spatial behaviour according to the specification of their correlation structure. Historically, one very good reason for concentrating on Gaussian models was that they are uniquely tractable as models for dependent data. With the increasing use of computationally intensive methods, and in particular of simulation-based methods of inference, the analytic tractability of Gaussian models is becoming a less compelling reason to use them. Nevertheless, it is still convenient to work within a standard model class in routine applications. The scope of the Gaussian model class can be extended by using a transformation of the original response variable, and with this extra flexibility the model often provides a good empirical fit to data. Also, within the specific context of geostatistics, the Gaussian assumption is the model-based counterpart of some widely used geostatistical prediction methods, including simple, ordinary and universal kriging (Journel and Huijbregts, 1978; Chilès and Delfiner, 1999). We shall use the Gaussian model initially as a model in its own right for geostatistical data with a continuously varying response, and later as an important component of a hierarchically specified generalised linear model for geostatistical data with a discrete response variable, as previously discussed in Section 1.4.

3.1 Covariance functions and the variogram

A *Gaussian spatial process*, $\{S(x) : x \in \mathbb{R}^2\}$, is a stochastic process with the property that for any collection of locations $x_1, \ldots, x_n$ with each $x_i \in \mathbb{R}^2$, the joint distribution of $S = \{S(x_1), \ldots, S(x_n)\}$ is multivariate Gaussian. Any process of this kind is completely specified by its *mean function*, $\mu(x) = \mathrm{E}[S(x)]$, and its *covariance function*, $\gamma(x, x') = \mathrm{Cov}\{S(x), S(x')\}$.

In any such process, consider an arbitrary set of locations $x_1, \ldots, x_n$, define $S = \{S(x_1), \ldots, S(x_n)\}$, write μ_S for the n-element vector with elements $\mu(x_i)$ and G for the $n \times n$ matrix with elements $G_{ij} = \gamma(x_i, x_j)$. Then, S follows a multivariate Gaussian distribution with mean vector μ_S and covariance matrix G. We write this as $S \sim \mathrm{MVN}(\mu_S, G)$.

Now, let $T = \sum_{i=1}^{n} a_i S(x_i)$. Then T is univariate Gaussian with mean $\mu_T = \sum_{i=1}^{n} a_i \mu(x_i)$ and variance

$$\sigma_T^2 = \sum_{i=1}^{n} \sum_{j=1}^{n} a_i a_j G_{ij} = a'Ga,$$

where $a = (a_1, \ldots, a_n)$. It must therefore be the case that $a'Ga \geq 0$. This condition, which must hold for all choices of n, $(x_1, \ldots, x_n)$ and $(a_1, \ldots, a_n)$ constrains G to be a *positive definite matrix*, and the corresponding $\gamma(\cdot)$ to be a *positive definite function*. Conversely, any positive definite function $\gamma(\cdot)$ is a legitimate covariance function for a spatial Gaussian process.

A spatial Gaussian process is *stationary* if $\mu(x) = \mu$, a constant for all x, and $\gamma(x, x') = \gamma(u)$, where $u = x - x'$ i.e., the covariance depends only on the vector difference between x and x'. Additionally, a stationary process is *isotropic* if $\gamma(u) = \gamma(||u||)$, where $||\cdot||$ denotes Euclidean distance i.e., the covariance between values of $S(x)$ at any two locations depends only on the distance between them. Note that the variance of a stationary process is a constant, $\sigma^2 = \gamma(0)$. We then define the *correlation function* to be $\rho(u) = \gamma(u)/\sigma^2$. The correlation function is symmetric in u i.e., $\rho(-u) = \rho(u)$. This follows from the fact that for any u, $\mathrm{Corr}\{S(x), S(x-u)\} = \mathrm{Corr}\{S(x-u), S(u)\} = \mathrm{Corr}\{S(x), S(x+u)\}$, the second equality following from the stationarity of $S(x)$. Hence, $\rho(u) = \rho(-u)$. From now on, we will use u to mean either the vector $x - x'$ or the scalar $||x - x'||$ according to context. We will also use the term *stationary* as a shorthand for stationary and isotropic. A process for which $S(x) - \mu(x)$ is stationary is called *covariance stationary*. Processes of this kind are very widely used in practice as models for geostatistical data.

In Chapter 2, we introduced the empirical variogram as a tool for exploratory data analysis. We now consider the theoretical variogram as an alternative characterisation of the second-order dependence in a spatial stochastic process. The *variogram* of a spatial stochastic process $S(x)$ is the function

$$V(x, x') = \frac{1}{2}\mathrm{Var}\{S(x) - S(x')\}. \tag{3.1}$$

Note that $V(x, x') = \frac{1}{2}[\mathrm{Var}\{S(x)\} + \mathrm{Var}\{S(x')\} - 2\mathrm{Cov}\{S(x), S(x')\}]$. In the stationary case, this simplifies to $V(u) = \sigma^2\{1 - \rho(u)\}$ which, incidentally, explains why the factor of one-half is conventionally included in the definition of the variogram. The variogram is also well defined as a function of u for a limited class of non-stationary processes; a one-dimensional example is a simple random walk, for which $V(u) = \alpha u$. Processes which are non-stationary but for which

$V(u)$ is well-defined are called *intrinsic random functions* (Matheron, 1973). We discuss these in more detail in Section 3.9.

In the stationary case the variogram is theoretically equivalent to the covariance function, but it has a number of advantages as a tool for data analysis, especially when the data locations form an irregular design. We discuss the data analytic role of the variogram in Chapter 5. Conditions for the theoretical validity of a specified class of variograms are usually discussed in terms of the corresponding family of covariance functions. Gneiting, Sasvári and Schlather (2001) present analogous results in terms of variograms.

3.2   Regularisation

In Section 1.2.1 we discussed briefly how the support of a geostatistical measurement could affect our choice of a model for the data. When the support for each measured value extends over an area, rather than being confined to a single point, the modelled signal $S(x)$ should strictly be represented as

$$S(x) = \int w(r)S^*(x - r)dr, \qquad (3.2)$$

where $S^*(\cdot)$ is an underlying, unobserved signal process and $w(\cdot)$ is a weighting function. In this case, the form of $w(\cdot)$ constrains the allowable form for the covariance function of $S(\cdot)$. Specifically, if $\gamma(\cdot)$ and $\gamma^*(\cdot)$ are the covariance functions of $S(\cdot)$ and $S^*(\cdot)$, respectively, it follows from (3.2) that

$$\gamma(u) = \int \int w(r)w(s)\gamma^*(u + r - s)drds. \qquad (3.3)$$

Now make a change of variable in (3.3) from s to $t = r - s$, and define

$$W(t) = \int w(r)w(t - r)dr.$$

Then (3.3) becomes

$$\gamma(u) = \int W(t)\gamma^*(u + t)dt. \qquad (3.4)$$

Typical weighting functions $w(r)$ would be radially symmetric, non-negative valued and non-increasing functions of $||r||$; this holds for the effect of the gamma camera integration in Example 1.3, where $w(r)$ is not known explicitly but is smoothly decreasing in $||r||$, and for the soil core data of Example 1.4, where $w(\cdot)$ is the indicator corresponding to the circular cross section of each core. In general, the effect of weighting functions of this kind is to make $S(x)$ vary more smoothly than $S^*(x)$, with a similar effect on $\gamma(u)$ by comparison with $\gamma^*(u)$.

An analogous result holds for the relationship between the variograms of $S(\cdot)$ and $S^*(\cdot)$. Using the relationship that $V(u) = \gamma(0) - \gamma(u)$ it follows from (3.4) that

$$V(u) = \int W(t)\{V^*(t + u) - V^*(t)\}dt. \qquad (3.5)$$

If the form of the weighting function $w(\cdot)$ is known, it would be possible to incorporate it into our model for the data. This would mean specifying a model for the covariance function of $S^\star(\cdot)$ and evaluating (3.4) to derive the corresponding covariance function of $S(\cdot)$. Note that this would enable data with different supports to be combined naturally, for example soil core data using different sizes of core. A more pragmatic strategy, and the only available one if $w(\cdot)$ is unknown, is to specify directly an appropriately smooth model for the covariance function of $S(\cdot)$.

The question of regularisation can also arise in connection with prediction, rather than model formulation. The simplest geostatistical prediction problem is to map the spatial signal $S(x)$, but in some applications a more relevant target for prediction might be a map of a regularised signal,

$$T(x) = \int S(u)du,$$

where the integral is over a disc with centre x i.e., $T(x)$ is a spatial average over the disc. We return to questions of this kind in Chapter 6.

3.3 Continuity and differentiability of stochastic processes

The specification of the covariance structure of a spatial process $S(x)$ directly affects the smoothness of the surfaces which the process generates. Accepted mathematical descriptors of the smoothness of a surface are its continuity and differentiability. However, for stochastically generated surfaces $S(x)$ we need to distinguish two kinds of continuity or differentiability. In what follows, we shall consider a one-dimensional space x, essentially for notational convenience.

We first consider *mean-square* properties, defined as follows. A stochastic process $S(x)$ is *mean-square continuous* if $\mathrm{E}[\{S(x+h) - S(x)\}^2] \to 0$ as $h \to 0$. Also, $S(x)$ is *mean-square differentiable*, with mean-square derivative $S'(x)$, if

$$\mathrm{E}\left[\left\{\frac{S(x+h) - S(x)}{h} - S'(x)\right\}^2\right] \to 0$$

as $h \to 0$. Higher-order mean-square differentiability is then defined sequentially in the obvious way; $S(x)$ is twice mean-square differentiable if $S'(x)$ is mean-square differentiable, and so on.

An important result, described for example in Bartlett (1955), is the following.

Theorem 3.1. A stationary stochastic process with correlation function $\rho(u)$ is k times mean-square differentiable if and only if $\rho(u)$ is $2k$ times differentiable at $u = 0$.

To examine differentiability at the origin of any particular correlation function $\rho(u)$, we need to consider the extended form of $\rho(u)$ in which u can take positive

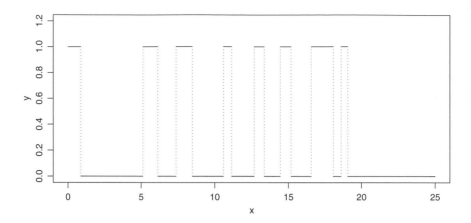

Figure 3.1. Realisation of a binary-values, mean-square continuous stochastic process (see text for details).

or negative arguments with $\rho(-u) = \rho(u)$. Hence, for example, the exponential correlation function $\rho(u) = \exp(-u/\phi)$ is continuous but not differentiable at the origin. In contrast, the Gaussian correlation function, defined by $\rho(u) = \exp\{-(u/\phi)^2\}$, is infinitely differentiable.

A second version of continuity and differentiability properties concerns *path continuity and differentiability*. A process $S(x)$ is *path-continuous*, or more generally k *times path-differentiable* if its realisations are continuous or k times differentiable functions, respectively.

In general, there need be no link between mean-square and path properties of stochastic processes. As a simple example, we can consider a binary-valued process $S(x)$ in which the real line is partitioned into a sequence of random intervals, whose lengths are independent realisations from a unit-mean exponential distribution, the value of $S(x)$ within each interval is zero with probability p, one otherwise, and the values of $S(x)$ on successive intervals are determined independently. Figure 3.1 shows a realisation with $p = 0.5$. Clearly, this process is not path-continuous. However, its correlation function is the exponential, $\rho(u) = \exp(-u)$, which is continuous at $u = 0$, hence $S(x)$ is mean-square continuous.

Kent (1989) gives a rigorous theoretical discussion of path-continuity for stationary, not necessarily Gaussian processes. Write $\rho(u) = p_m(u) + r_m(u)$, where $p_m(u)$ is the polynomial of degree m given by the Taylor series expansion of $\rho(u)$ about $u = 0$. Then, a sufficient condition for the existence of a path-continuous two-dimensional stationary process with correlation function $\rho(\cdot)$ is that $\rho(\cdot)$ is twice continuously differentiable and $|r_2(u)| = O(u^2/|\log u|^{3+\gamma})$ as $u \to 0$, for some $\gamma > 0$. A slightly stronger condition which is easier to check in practice is that $|r_2(u)| = O(u^{2+\epsilon})$ for some $\epsilon > 0$. For stationary Gaussian processes in two dimensions, a sufficient condition for path-continuity is that $\rho(0) - \rho(u) = O(1/|\log u|^{1+\epsilon})$, which is only slightly stronger than the requirement for mean-square continuity, namely that $\rho(\cdot)$ is continuous at the origin.

This justifies using mean-square differentiability as a convenient measure of the smoothness of stationary Gaussian processes when considering their suitability as empirical models for natural phenomena.

3.4 Families of covariance functions and their properties

Positive definiteness is the necessary and sufficient condition for a parametric family of functions to define a legitimate class of covariance functions, but this is not an easy condition to check directly. For this reason, it is useful to have available a range of standard families which are known to be positive definite but in other respects are sufficiently flexible to meet the needs of applications to geostatistical data. In this section, we give the details of several such families and outline their properties. Our concern here is with models for processes in two spatial dimensions. All of the covariance families which we describe are also valid in one or three dimensions. In general, a valid covariance family in $\mathbb{R}^d$ does not necessarily remain valid in more than d spatial dimensions, but is automatically valid in dimensions less than d.

3.4.1   The Matérn family

The most common form of empirical behaviour for stationary covariance structure is that the correlation between $S(x)$ and $S(x')$ decreases as the distance $u = ||x - x'||$ increases. It is therefore natural to look for models whose theoretical correlation structure behaves in this way. In addition, we can expect that different applications may exhibit different degrees of smoothness in the underlying spatial process $S(x)$.

The Matérn family of correlation functions, named after Matérn (1960), meets both of these requirements. It is a two-parameter family,

$$\rho(u) = \{2^{\kappa-1}\Gamma(\kappa)\}^{-1}(u/\phi)^\kappa K_\kappa(u/\phi), \tag{3.6}$$

in which $K_\kappa(\cdot)$ denotes a modified Bessel function of order κ, $\phi > 0$ is a scale parameter with the dimensions of distance, and $\kappa > 0$, called the *order*, is a shape parameter which determines the analytic smoothness of the underlying process $S(x)$. Specifically, $S(x)$ is $\lceil \kappa-1 \rceil$ times mean-square differentiable, where $\lceil \kappa \rceil$ denotes the smallest integer greater than or equal to κ

Figure 3.2 shows the Matérn correlation function for each of $\kappa = 0.5$, 1.5 and 2.5, corresponding to processes $S(x)$ which are mean-square continuous, once differentiable and twice differentiable, respectively. In the diagram, the values of ϕ have been adjusted so as to give all three functions the same *practical range*, which we define here as the distance u at which the correlation is 0.05. For Figure 3.2 we used $u = 0.75$ as the value of the practical range. For $\kappa = 0.5$, the Matérn correlation function reduces to the exponential, $\rho(u) = \exp(-u/\phi)$, whilst as $\kappa \to \infty$, $\rho(u) \to \exp\{-(u/\phi)^2\}$ which is also called the Gaussian correlation function or, somewhat confusingly in the present context, the Gaus-

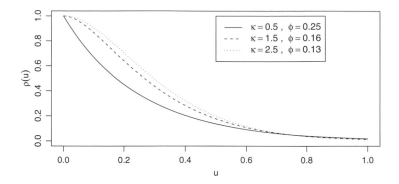

Figure 3.2. Matérn correlation functions with, $\kappa = 0.5$ (solid line), $\kappa = 1.5$ (dashed line) and $\kappa = 2.5$ (dotted line), and adjusted values of ϕ for equivalent practical ranges.

sian model. Whittle (1954) proposed the special case of the Matérn correlation function with $\kappa = 1$.

Note that the parameters ϕ and κ in (3.6) are non-orthogonal, in the following sense. If the true correlation structure is Matérn with parameters ϕ and κ, then the best-fitting approximation with order $\kappa^* \neq \kappa$ will also have $\phi^* \neq \phi$. In other words, scale parameters corresponding to different orders of Matérn correlation are not directly comparable. The relationship between the practical range and the scale parameter ϕ therefore depends on the value of κ. For instance, the practical range as defined above is approximately 3ϕ, 4.75ϕ and 5.92ϕ for the Matérn functions with $\kappa = 0.5$, 1.5 and 2.5, respectively, and $\sqrt{3}\phi$ for the Gaussian correlation function. For this reason, Handcock and Wallis (1994) suggest a re-parametrisation of (3.6) from κ and ϕ to a more nearly orthogonal pair κ and $\alpha = 2\phi\sqrt{\kappa}$. The re-parametrisation does not, of course, change the model but is relevant to our discussion of parameter estimation in Chapters 5 and 7.

Figure 3.3 shows a one-dimensional trace through a simulated realisation of a spatial Gaussian process with each of the Matérn correlation functions above, using the same random seed for all three realisations. The increasing analytic smoothness of the process as κ increases is reflected in the visual appearance of the three realisations, but the more noticeable difference is between the non-differentiable and the differentiable case i.e., between $\kappa = 0.5$ on the one hand and $\kappa = 1.5$ or $\kappa = 2.5$ on the other.

Figure 3.4 shows simulated two-dimensional realisations of Gaussian processes whose correlation functions are Matérn with $\kappa = 0.5$ and $\kappa = 2.5$, again using the same random number seed to make the realisations directly comparable. The difference in smoothness between the non-differentiable and differentiable cases is again visually striking.

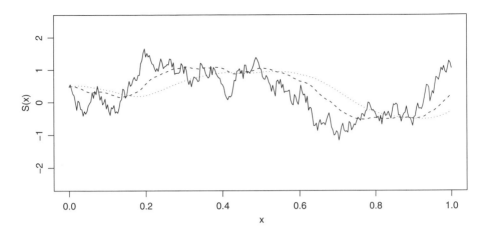

Figure 3.3. One-dimensional realisations of spatial Gaussian processes whose correlation functions are Matérn with $\kappa = 0.5$ (solid line), $\kappa = 1.5$ (dashed line) and $\kappa = 2.5$ (dotted line).

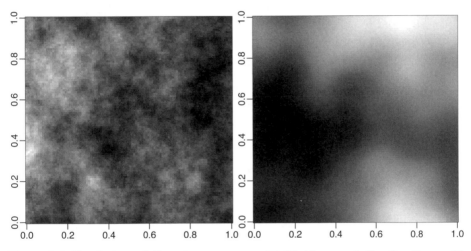

Figure 3.4. Simulations of Gaussian processes with Matérn correlation functions with $\kappa = 0.5$ and $\phi = 0.25$ (left) and $\kappa = 2.5$ and $\phi = 0.13$ (right).

3.4.2 The powered exponential family

This family is defined by the correlation function

$$\rho(u) = \exp\{-(u/\phi)^{\kappa}\}. \tag{3.7}$$

Like the Matérn family, it has a scale parameter $\phi > 0$, a shape parameter κ, in this case bounded by $0 < \kappa \leq 2$, and generates correlation functions which are monotone decreasing in u. Also like the Matérn family the relation between the practical range and the parameter ϕ will depend on the value of κ. However, the family is less flexible than the Matérn, in the sense that the underlying Gaussian process $S(x)$ is mean-square continuous and not mean-

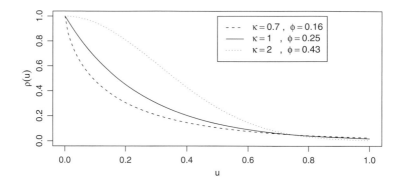

Figure 3.5. Powered exponential correlation functions with $\kappa = 0.7$ (dashed line), $\kappa = 1$ (solid line) and $\kappa = 2$ (dotted line) and values of *phi* adjusted such that the practical range is 0.75.

square differentiable for all $0 < \kappa < 2$ but infinitely mean square differentiable when $\kappa = 2$, the maximum legitimate value. Figure 3.5 shows the powered exponential correlation function for each of $\kappa = 0.7$, 1 and 2, and with values of ϕ adjusted to provide the same practical range of 0.75. Figure 3.6 shows one-dimensional realisations of the corresponding Gaussian processes $S(x)$. We used the same seed as for the earlier simulations of the Matérn model. The realisation for the powered exponential model with $\kappa = 1$ is therefore the same as for the Matérn model with $\kappa = 0.5$. Notice that the realisations for $\kappa = 0.7$ and $\kappa = 1$, both of which correspond to mean-square continuous but non-differentiable processes, look rather similar in character.

The extreme case $\kappa = 2$, which is equivalent to the limiting case of a Matérn correlation function as $\kappa \to \infty$, can generate very ill-conditioned covariance structure. A process $S(x)$ with this correlation function has the theoretical property that its realisation on an arbitrarily small, continuous interval determines the realisation on the whole real line. For most applications, this would be considered unrealistic.

3.4.3 Other families

In classical geostatistics, the *spherical family* is widely used. This has correlation function

$$\rho(u) = \begin{cases} 1 - \frac{3}{2}(u/\phi) + \frac{1}{2}(u/\phi)^3 & : \quad 0 \le u \le \phi \\ 0 & : \quad u > \phi \end{cases} \tag{3.8}$$

where $\phi > 0$ is a single parameter with the dimensions of distance. One qualitative difference between this and the families described earlier is that it has a finite range i.e., $\rho(u) = 0$ for sufficiently large u, namely $u > \phi$. The spherical family lacks flexibility by comparison with the two-parameter Matérn class. Also, $\rho(u)$ is only once differentiable at $u = \phi$, which causes technical difficulties with maximum likelihood estimation (Warnes and Ripley, 1987; Mardia and Watkins, 1989). The left-hand panel in Figure 3.7 shows the spherical cor-

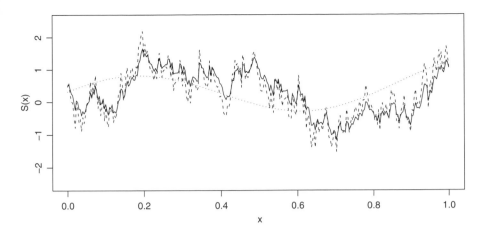

Figure 3.6. One-dimensional realisations of spatial Gaussian processes whose correlation functions are powered exponential, $\kappa = 0.7$ (dashed line), $\kappa = 1$ (solid line) and $\kappa = 2$ (dotted line).

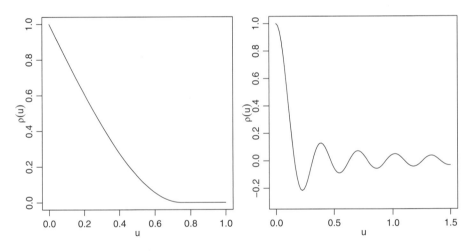

Figure 3.7. Correlation functions, the spherical (left) with $\phi = 0.75$ and wave (right) with $\phi = 0.05$.

relation function when $\phi = 0.75$. The corresponding Gaussian process $S(x)$ is mean-square continuous but non-differentiable. The name and algebraic form of the spherical family derives from the geometry of intersecting spheres; see Exercise 3.3.

Non-monotone correlation functions are rare in practice. One example of a valid non-monotone family is

$$\rho(u) = (u/\phi)^{-1} \sin(u/\phi) \tag{3.9}$$

where $\phi > 0$ is a single parameter, again with the dimension of distance. The right-hand panel of Figure 3.7 illustrates the characteristic damped oscillatory

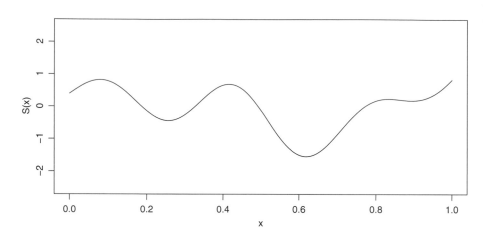

Figure 3.8. One-dimensional realisation of a spatial Gaussian process whose correlation function is $\rho(u) = (u/0.05)^{-1}\sin(u/0.05)$.

behaviour of this correlation function, whilst Figure 3.8 shows a realisation of the corresponding process $S(x)$. Notice how the oscillatory nature of the correlation function is reflected in the oscillatory behaviour of the simulated realisation.

Other classes of correlation function, and criteria to check the validity of candidate functions, are described in Schlather (1999), who in turn draws on material in Gneiting (1997). However, for most geostatistical applications the families described here should be sufficient, if only because more elaborate models are hard to identify unless the available data are abundant. In general, we favour the Matérn family because of its flexibility, coupled with the tangible interpretation of the shape parameter κ as a measure of the differentiability of the underlying process $S(x)$. Also, because of the difficulty of identifying all the parameters of this model empirically, we would usually either fix the value of κ according to the context of the application, or choose amongst a limited set of values of κ, for example $\kappa = 0.5, 1.5, 2.5$ as illustrated in Figure 3.2.

3.5   The nugget effect

In geostatistical practice, the term "nugget effect" refers to a discontinuity at the origin in the variogram. Within our model-based framework, its literal interpretation is as the measurement error variance, τ^2, or equivalently the conditional variance of each measured value Y_i given the underlying signal value $S(x_i)$. Formally, this amounts to modelling the measurement process, $Y(x)$ say, as a Gaussian process whose correlation function is discontinuous at the origin, hence

$$\text{Corr}\{Y(x), Y(x')\} = \begin{cases} 1 & : \quad x = x' \\ \sigma^2\rho(||x - x'||)/(\sigma^2 + \tau^2) & : \quad x \neq x' \end{cases}$$

where $\rho(\cdot)$ is the (continuous) correlation function of $S(x)$ and $\|\cdot\|$ denotes distance.

In practice, when the sampling design specifies a single measurement at each of n distinct locations, the nugget effect has a dual interpretation as either measurement error or spatial variation on a scale smaller than the smallest distance between any two points in the sample design, or any combination of these two effects. These two components of the nugget effect can only be separately identified if the measurement error variance is either known, or can be estimated directly using repeated measurements taken at coincident locations.

3.6 Spatial trends

The simplest form of departure from stationarity is to allow the mean response, $\mu(x)$, to depend on location. We call any such varying mean a *spatial trend*. In applications, we may choose to model $\mu(x)$ directly as a function of x. In practice, this is most often done through a polynomial regression model, using powers and cross products of the Cartesian coordinates of x as explanatory variables. Models of this kind are called *trend surface* models. They rarely have any scientific foundation. Our view is that linear or quadratic trend surfaces can provide useful empirical descriptions of simple, unexplained spatial trends, but that higher-degree surfaces should be avoided because complicated trends are better described through the stochastic component of the model. See, for example, our illustrative analysis of the surface elevation data reported in Chapter 2.

A more interesting kind of spatial trend arises when the mean function can be modelled using spatially referenced covariates, hence for example $\mu(x) = \alpha + d(x)\beta$ where $d(x)$ is a scientifically relevant property of the location x. In our opinion, models of this kind are more interesting than trend surface models because they seek to explain, rather than merely to describe, the spatial variation in the response variable. For example, in the Gambia malaria data of Example 1.3 modelling the spatial variation in prevalence as a function of greenness has a natural scientific interpretation because the greenness index is a surrogate measure of the suitability of each location for mosquitos to breed. If, hypothetically, greenness showed a smooth east-west trend, then modelling malaria prevalence as a function of greenness or as a function of longitude might give equally good empirical fits to the data, but modelling prevalence as a function of greenness would offer the more satisfying explanation and would be the more likely to translate to other study regions.

As discussed in Section 1.2.2, when values of a potential explanatory variable $d(x)$ are only recorded at the same locations as give rise to the basic geostatistical data (x_i, y_i), we need to consider whether we should treat $d(x)$ as a second, stochastic variable to be analysed jointly with the primary signal process, $S(x)$, rather than as a deterministic quantity.

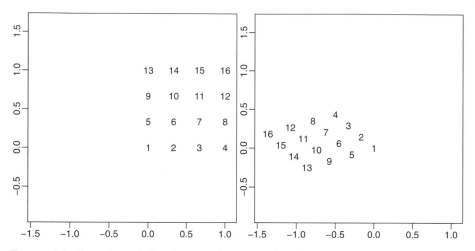

Figure 3.9. Rotation of the data configuration by the anisotropy parameters. The left-hand panel shows the original locations, the right-hand panel the transformed locations in isotropic space when $\psi_A = 2\pi/3$ and $\psi_R = 2$.

3.7   Directional effects

Another form of non-stationarity is non-stationarity in the covariance structure. One specific way to relax the stationarity assumption is to allow directional effects so that, for example, the rate at which the correlation decays with increasing distance is allowed to depend also on the relative orientation between pairs of locations.

The simplest form of directional effect on the covariance structure is called *geometrical anisotropy*. This arises when a stationary covariance structure is transformed by a differential stretching and rotation of the coordinate axes. Hence, geometrical anisotropy is defined by two additional parameters. Algebraically, a model with geometrical anisotropy in spatial coordinates $x = (x_1, x_2)$ can be converted to a stationary model in coordinates $x' = (x'_1, x'_2)$ by the transformation

$$(x'_1, x'_2) = (x_1, x_2) \begin{bmatrix} \cos(\psi_A) & -\sin(\psi_A) \\ \sin(\psi_A) & \cos(\psi_A) \end{bmatrix} \begin{bmatrix} 1 & 0 \\ 0 & \psi_R^{-1} \end{bmatrix} \qquad (3.10)$$

where ψ_A is called the *anisotropy angle* and $\psi_R > 1$ is called the *anisotropy ratio*. The direction along which the correlation decays most slowly with increasing distance is called the *principal axis*.

These operations are illustrated in Figure 3.9. The original locations are shown in the left-hand panel. Suppose that the anisotropy angle is $\psi_A = 2\pi/3$, and the anisotropy ratio is $\psi_R = 2$. Then, applying the coordinate transformation (3.10) we obtain the locations in the right-hand panel, which are now in an isotropic space, and proceed to fit an isotropic model in this transformed space. In practice, ψ_A and ψ_R are unknown, and the model fit would be optimised by treating ψ_A and ψ_R as additional parameters to be estimated.

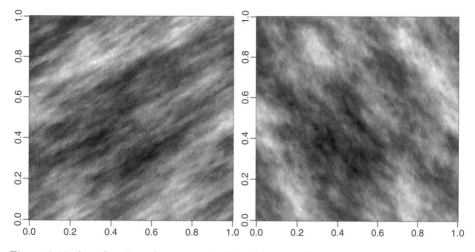

Figure 3.10. A realisation of a geometrically anisotropic Gaussian spatial process whose principal axis runs diagonally across the square region with anisotropy parameters $(\pi/3, 4)$ for the left-hand panel and $(3\pi/4, 2)$ for the right-hand panel.

Figure 3.10 shows realisations of two Gaussian spatial process with geometrical anisotropy. The directional effects are visually clear, with the principal axis in each case running diagonally over the square region shown. For the left panel the anisotropy angle is $\pi/3$ radians and the anisotropy ratio is 4. For the right panel the anisotropy angle is $3\pi/4$ radians and the anisotropy ratio is 2. The two processes have common parameter values $\mu = 0$, $\sigma^2 = 1$ and exponential correlation function with $\phi = 0.25$, and the two realisations were generated using the same random seed.

Note that geometric anisotropy cannot describe local directional features of a spatial surface, only global ones. On the other hand, the presence of local directional features in a realisation of a spatial process need not imply that the underlying process is anisotropic. Consider, for example, a surface constructed as the superposition of profiles $f(\cdot)$ translated by the points of a homogeneous Poisson point process. Thus,

$$S(x) = \sum_{i=1}^{\infty} f(x - X_i) \tag{3.11}$$

where the X_i are the points of the Poisson process. Figure 3.11 compares realisations of two such processes in which the intensity of the Poisson process is 16 points per unit area and the profile function is the probability density of a bivariate Gaussian distribution with zero mean, standard deviation 0.1 in each coordinate direction and correlation 0.75. In the left-hand panel, the global directional feature along the diagonal direction is clear. In the right-hand panel, each profile has been randomly rotated so that, whilst local directional effects can still be seen, the resulting model is isotropic with no global directional effects. Higdon (1998, 2002) has proposed constructions similar to, but more general than, (3.11) to define a general class of non-stationary, non-Gaussian models.

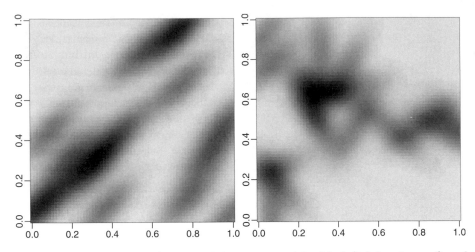

Figure 3.11. Realisations of two spatial processes with global (left-hand panel) and local (right-hand panel) directional effects. See text for detailed explanation.

Geometrical anisotropy deals with a particular form of non-stationarity by transforming the space x using stretching and rotation, so that the underlying process is stationary in the transformed space. Sampson and Guttorp (1992), Guttorp, Meiring and Sampson (1994) and Guttorp and Sampson (1994) develop a more general version of this approach. Their method seeks a smooth deformation of the x-space, equivalent to a transformation from x to x^* say, so that the covariance function depends only on distance in the deformed space, hence for any two locations x and y in the original space, $\mathrm{Cov}\{S(x), S(y)\} = \gamma(||x^* - y^*||)$. Perrin and Meiring (1999) discuss identifiability issues for this class of models, whilst Schmidt and O'Hagan (2003) develop a Bayesian version. Replicated observations are needed at each sampling location in order to identify the required transformation. In practice, the approach is feasible when a time series is collected at each location as this gives the necessary, albeit dependent, replication.

Non-stationarity can also arise because Euclidean distance is not an appropriate measure of spatial separation. For example, Rathbun (1998) considers non-Euclidean distances in modelling spatial variation in an estuary where, amongst other considerations, the line segment joining two locations within the estuary may cross a stretch of land.

3.8   Transformed Gaussian models

We now expand the discussion of Section 2.2, where we mentioned briefly that the range of applicability of the Gaussian model can be extended by assuming that the model holds after a marginal transformation of the response variable.

As in other areas of statistics, there are at least three different reasons for using a transformation of the data. Firstly, a particular transformation might be suggested by qualitative arguments, or even by convention. For example, if effects are thought to be operating multiplicatively, then a log-transformation converts the problem to a scale on which effects are, more conveniently, additive. Secondly, a transformation may be used as a variance-stabilising device for a known, non-Gaussian sampling distribution. For example, square root and arc-sine transformations approximately stabilise the sampling variance under Poisson and binomial sampling, respectively. Note, however, that there is no reason why a transformation which stabilises the variability in the measurements conditional on the signal should also stabilise the variability in the signal, or *vice versa*. The transformation approach to variance instability used to be widespread in regression modelling of non-Gaussian data, but has largely been replaced by the use of generalized linear models (McCullagh and Nelder, 1989). Section 1.4 and, in more detail, Chapter 4 describe an extension of classical generalized linear models to accommodate non-Gaussian geostatistical data. Finally, we can introduce a parametric family of transformations simply as an empirical generalisation of the Gaussian model, in which case the choice of a particular transformation corresponds to the estimation of an additional parameter. The most widely used example of this approach is the Box-Cox family of transformations (Box and Cox, 1964),

$$Y^* = \begin{cases} (Y^\lambda - 1)/\lambda & : \quad \lambda \neq 0 \\ \log Y & : \quad \lambda = 0. \end{cases} \tag{3.12}$$

The log-transformation is perhaps the most widely used in practice, and explicit expressions can be derived for its mean and covariance structure. Suppose that $T(x) = \exp\{S(x)\}$, where $S(x)$ is a stationary Gaussian process with mean μ, variance σ^2 and correlation function $\rho(u)$. The moment generating function of $S(x)$ is

$$M(a) = \mathrm{E}[\exp\{aS(x)\}] = \exp\{a\mu + \frac{1}{2}a^2\sigma^2\}. \tag{3.13}$$

It follows from (3.13), setting $a = 1$, that $T(x)$ has expectation

$$\mu_T = \exp\left(\mu + \frac{1}{2}\sigma^2\right). \tag{3.14}$$

Similarly, setting $a = 2$ in (3.13) gives $\mathrm{E}[T(x)^2]$, and hence the variance of $T(x)$ as

$$\sigma_T^2 = \exp(2\mu + \sigma^2)\{\exp(\sigma^2) - 1\}. \tag{3.15}$$

Finally, for any two locations x and x', $T(x)T(x') = \exp\{S(x) + S(x')\}$, and $S(x) + S(x')$ is Gaussian with mean $m = 2\mu$ and variance $v = 2\sigma^2\{1 + \rho(||x - x'||)\}$. It follows that $\mathrm{E}[T(x)T(x')] = \exp(m + v/2)$, and straightforward algebra gives the correlation function of $T(x)$ as

$$\rho_T(u) = [\exp\{\sigma^2\rho(u)\} - 1]/[\exp\{\sigma^2\} - 1]. \tag{3.16}$$

Note that the mean and variance of $T(x)$ depend on both μ and σ^2, whereas the correlation function of $T(x)$ does not depend on μ.

Log-Gaussian processes exhibit, to a greater or lesser extent depending on the values of the model parameters, asymmetric behaviour with local patches

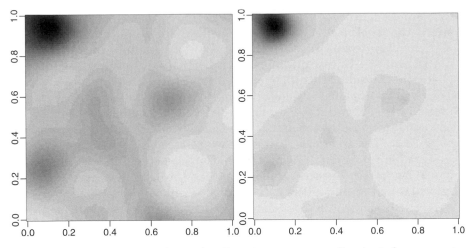

Figure 3.12. Realisations of two log-Gaussian processes. See text for parameter specifications.

of values close to zero, interspersed with relatively sharp peaks. In particular, we can write any Gaussian process $S(x)$ as $\mu + \sigma Z(x)$, and the corresponding log-Gaussian process as $T(x) = \alpha T_0(x)^\sigma$, where $\alpha = \exp(\mu)$ and $T_0(x) = \exp\{Z(x)\}$. Hence, for any given $Z(x)$, the value of μ affects the scale of the surface $T(x)$, whilst σ affects its shape, with larger values of σ producing sharper peaks and flatter troughs

The two panels of Figure 3.12 illustrate this affect. They show realisations of two log-Gaussian processes of the form $T(x) = \exp\{\sigma Z(x)\}$, where $Z(x)$ is a Gaussian process with zero mean, unit variance and Matérn correlation of order $\kappa = 1.5$ and with range parameter $\phi = 0.2$. Both panels use the same realisation of $Z(x)$ and differ only in that the left-hand panel has $\sigma = 0.1$ and the right-hand panel $\sigma = 0.7$.

The two panels of Figure 3.13 compare a realisation of a log-Gaussian process and a Gaussian process with the same mean and variance, and closely matched correlation structure. The log-Gaussian process used for the left-hand panel of Figure 3.13 has its correlation structure $\rho_T(u)$ induced by an underlying Matérn correlation function $\rho_0(u)$ with parameters $\kappa = 1.5$ and $\phi = 0.2$, and variance $\sigma^2 = 1$. We then used a simple least squares criterion to obtain a Matérn correlation function, $\rho_A(u)$ say, which approximated $\rho_T(u)$ as closely as possible, resulting in the parameter values $\phi_a = 0.18$ and $\kappa_a = 1.32$. To obtain the right-hand panel of Figure 3.13 we then simulated a Gaussian process using the correlation function $\rho_A(u)$ in conjunction with a mean and variance chosen so as to match those of the log-Gaussian process. As usual, we used the same random number seed for the two realisations being compared. Figure 3.14 compares the correlation functions $\rho_T(u)$, $\rho_A(u)$ and $\rho_0(u)$. We see that the correlation functions of the processes used to generate the two realisations shown in Figure 3.13 are almost identical, yet the realisations themselves are very different in character because of their different distributional properties.

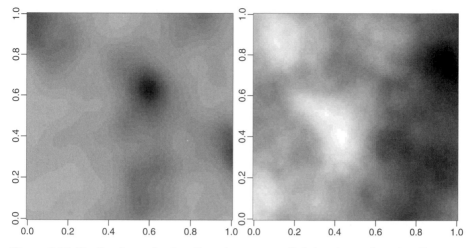

Figure 3.13. Realisations of a log-Gaussian process (left-hand panel) and a Gaussian process with closely matched correlation structure (right-hand panel). See text for parametric specifications.

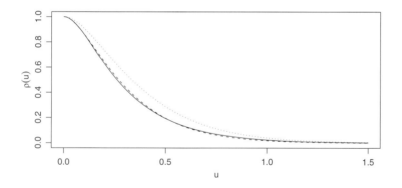

Figure 3.14. Correlation functions $\rho_T(u)$ (solid line) and $\rho_A(u)$ (dashed line) for the log-Gaussian and Gaussian processes whose realisations are compared in Figure 3.13. The dotted line shows the Matérn correlation function $\rho_0(u)$. See text for parametric specifications.

3.9 Intrinsic models

In Section 3.6 we discussed a simple form of non-stationary model, namely the sum of a deterministic spatial trend and a stochastic, spatially correlated residual. Similarly, in Section 3.7 we discussed a deterministic strategy for dealing with non-stationarity, in this case a transformation of the spatial coordinate system to deal with a global directional effect in the underlying process. An alternative strategy is to treat non-stationarity as an inherently stochastic phenomenon.

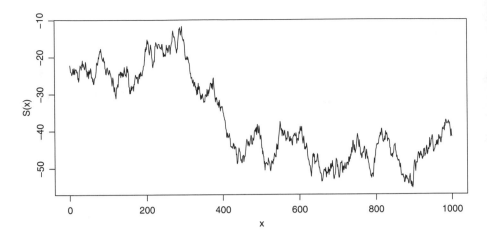

Figure 3.15. Realisation of a one-dimensional random walk. See text for detailed explanation

As a simple, spatially discrete one-dimensional example of an intrinsic model we consider a random walk, $S(x)$, defined recursively by

$$S(x) = S(x-1) + Z(x) : x = 0, 1, \ldots \qquad (3.17)$$

where the $Z(x)$ are mutually independent, normally distributed with mean 0 and variance 1. Conventionally, we add the initial condition that $S(0) = 0$, in which case $\mathrm{E}[S(x)] = 0$ for all x and $\mathrm{Var}\{S(x)\} = x$. However, an alternative interpretation, which is perhaps more natural in the spatial setting where any ordering along the coordinate axes is arbitrary, is that $S(x)$ fluctuates randomly about an arbitrary level i.e., the average is indeterminate and the variation observed within a finite segment of space increases with the length of the segment.

Figure 3.15 shows a simulated realisation of such a process. The process was initialised at zero, allowed to run for 5000 steps, then observed and plotted for an additional 1000 steps. By chance, the plotted values of $S(x)$ vary over the approximate range -60 to -10 although their theoretical expectation over repeated realisations is zero. More interestingly, the initial and final portions of Figure 3.15 appear on casual inspection to be approximately stationary whereas, the portion between $x = 300$ and $x = 450$ suggests a decreasing, approximately linear trend. One lesson which we take from this example is that when our data consist of a single realisation of a correlated stochastic process, it is often the case that qualitatively wrong models can give a reasonable empirical fit to the data.

The random walk model (3.17) is an example of a general class of non-stationary stochastic processes known as *intrinsic random functions* (Matheron, 1973). An intrinsic random function is a stochastic process $S(x)$ with stationary increments. This means that for any $u \in \mathbb{R}^2$, the process $D_u(x)$ defined by

$$D_u(x) = S(x) - S(x-u)$$

is stationary. Suppose that $\mathrm{Var}(D_u) = \sigma_u^2$. Then, $\frac{1}{2}\sigma_u^2$, regarded as a function of u, is also the variogram of $S(x)$. Hence, intrinsic random functions can be thought of as processes for which the variogram, but not necessarily the covariance function, depends only on u. For the random walk process (3.17), the variogram is $V(u) = \frac{1}{2}\mathrm{Var}\{S(x) - S(x-u)\} = \frac{1}{2}u$, for $u \geq 0$, whereas the covariance function is $\gamma(x, u) = \mathrm{Cov}\{S(x), S(x-u)\} = |x - u|$, which depends on both u and x.

Examples of legitimate intrinsic variogram models include power law and logarithmic forms. The power law model, $V(u) = (u/\phi)^\kappa$ is valid for $0 < \kappa < 2$. The most widely used special case is the linear variogram, $V(u) = u/\phi$. The logarithmic model,

$$V(u) = \log(u/\phi), \tag{3.18}$$

occupies a special place in classical geostatistics because of its connection to an empirical law discovered by De Wijs (1951, 1953). De Wijs observed that when a sample of ore was broken into smaller pieces, the variability between the grades of the pieces in relation to the average grade of the original sample appeared to depend only on the ratio of the volume of the pieces to the volume of the original, and not on the absolute volume of the original. Viewed as a model for a variogram, (3.18) has the unattractive property that $V(u) \to -\infty$ as $u \to 0$ which is incompatible with the definition of the variogram as a variance. However, suppose that (3.18) holds for an unobserved process $S^*(x)$, and that we observe

$$S(x) = \int w(r)S^*(x - r)dr, \tag{3.19}$$

where $w(u)$ is a non-negative valued weighting function. As discussed in Section 1.2.1 this corresponds to each observed measurement having a finite support deriving from a finite spatial neighbourhood centred on the point x. Now, as in the derivation of (3.5), write

$$W(t) = \int w(r)w(t - r)dr.$$

Combining (3.18) and (3.5) then gives the variogram of the regularised process as

$$\begin{aligned} V(u) &= \int W(t)[\log\{(t + u)/\phi\} - \log(t/\phi)]dt \\ &= \int W(t)\{\log(t + u) - \log(t)\}dt, \end{aligned} \tag{3.20}$$

which is non-negative valued for all $u \geq 0$ and does not depend on ϕ. This rather surprising result is the theoretical analogue of De Wijs's empirical law. Besag and Mondal (2005) establish a close theoretical link between the De Wijs process and intrinsic autoregressive processes on a two-dimensional lattice and show that, by making the lattice spacing sufficiently fine, the spatially discrete autoregressive process can give an excellent approximation to the spatially continuous De Wijs process. The lattice formulation also brings substantial computational benefits for large data-sets.

Intrinsic random functions embrace a wider class of models than do stationary random functions. With regard to spatial prediction, the main difference between predictions obtained from intrinsic and from stationary models is that if intrinsic models are used, the prediction at a point x is influenced by the local behaviour of the data i.e., by the observed measurements at locations relatively close to x, whereas predictions from stationary models are also affected by global behaviour. One way to understand this is to remember that the mean of an intrinsic process is indeterminate. As a consequence, predictions derived from an assumed intrinsic model tend to fluctuate around a local average. In contrast, predictions derived from an assumed stationary model tend to revert to the global mean of the assumed model in areas where the data are sparse. Which of these two types of behaviour is the more natural depends on the scientific context in which the models are being used.

3.10   Unconditional and conditional simulation

Simulation plays an important role in geostatistical practice, both in conducting Monte Carlo experiments to gain insight into the properties of particular models and associated statistical methods, and as a fundamental tool in conducting geostatistical inference when the required analytical results are intractable.

The most basic simulation problem is to simulate a realisation, say $Y = (Y_1, \ldots, Y_n)$, of a Gaussian model at a set of n locations $x_i \in \mathbb{R}^2$. Note firstly that if the model for Y includes a nugget effect, with nugget variance τ^2, we can represent Y as $Y = \mu + S + \tau T$ where $\mu = \mathrm{E}[Y]$, $T = (T_1, .., T_n)$ is a set of mutually independent $\mathrm{N}(0, 1)$ random variables, and the spatial signal $S = (S_1, \ldots, S_n)$ follows a zero-mean multivariate Gaussian distribution, namely $S \sim \mathrm{MVN}(0, \Sigma)$.

The standard method for simulating a realisation of S is to simulate an independent random sample $Z = (Z_1, \ldots, Z_n)$ from the standard Gaussian distribution, $\mathrm{N}(0, 1)$, and apply a linear transformation,

$$S = AZ, \tag{3.21}$$

where A is any matrix such that $AA' = \Sigma$. Two ways to construct A are through Cholesky factorisation and singular value decomposition.

The Cholesky factorisation of Σ is $\Sigma = LL'$, where L is a lower-triangular matrix. Hence in (3.21) we take $A = L$. Because A is lower triangular, this method of simulating S can be interpreted as first simulating S_1 from its marginal, univariate Gaussian distribution, then successively simulating $S_2, \ldots, S_n$ from the conditional distributions of each S_i given $S_1, \ldots, S_{i-1}$, each of which is again univariate Gaussian.

The singular value decomposition of Σ is $\Sigma = U\Lambda U'$, where Λ is a diagonal matrix whose diagonal elements $\lambda = (\lambda_1, \ldots, \lambda_n)$ are the eigenvalues of Σ, ordered from largest to smallest, whilst the columns of U contain the corresponding eigenvectors, hence $U'U = I$. Because Σ is positive definite, all of the

λ_i are positive. Hence, a second possible choice for A in (3.21) is $A = U\Lambda^{\frac{1}{2}}$, where $\Lambda^{\frac{1}{2}}$ is the diagonal matrix with diagonal elements $\sqrt{\lambda_i}$.

Simulating realisations of the stationary Gaussian model by either of these methods becomes difficult in practice when n is very large, because of the computational burden associated with the necessary matrix operations. Typically, to simulate a realisation of a process $S(\cdot)$ over a spatial region, A say, we would approximate the spatially continuous surface $S(x)$ by its values on a fine grid to cover the region of interest. For this situation, Wood and Chan (1994) provide an ingenious algorithm which uses circulant embedding in conjunction with fast Fourier transform methods to achieve very substantial reductions in both computing time and storage requirements when the number of grid points is large; for example, simulation on a grid of size 256 by 256 becomes computationally straightforward.

A completely different approach is to use a Markov chain Monte Carlo method known as Gibbs sampling (Gilks et al., 1996). Define the *full conditional distributions* of $S = (S_1, \ldots, S_n)$ as the n univariate Gaussian distributions of each S_i given all other S_j. Choose any initial set of values for S, say $S_0 = (S_{01}, \ldots, S_{0n})$. Now, simulate a new set of values, $S_1 = (S_{11}, \ldots, S_{1n})$ successively from the full conditionals of each S_i given the *new* values $S_{1j} : j = 1, \ldots, i-1$ and the *old* values $S_{0j} : j = i+1, \ldots, n$, with the obvious interpretations for $i = 1$ and $i = n$. This defines a single *sweep* of the Gibbs sampler. Re-set S_0 to be the newly simulated S_1 and repeat. If we iterate this process over many sweeps, the distribution of the resulting sequence of simulations S_1 converges to the required multivariate Gaussian.

For the models considered in this chapter, the Gibbs sampler is generally not a sensible option because the evaluation of each full conditional distribution requires the inversion of an $(n-1) \times (n-1)$ covariance matrix. However, the method becomes very attractive if we *define* our models by the form of their full conditionals, especially so if the full conditionals are sparse i.e., the full conditional of each S_i depends only on a small number of S_j, called the *neighbours* of S_i. Models of this kind are known as Gaussian Markov random fields and are discussed in Rue and Held (2005). For general geostatistical applications, Markov random field models have the unattractive feature that they are tied to a specified set of locations rather than being defined in a spatially continuous way. Hence, they cannot be used directly to make spatially continuous predictions. However, Rue and Tjelmeland (2002) have shown how a spatially continuous Gaussian process can be approximated by a Gaussian Markov random field on a fine grid. Hence, a feasible strategy is to define a spatially continuous model but use its approximating Markov random field for computation.

In the geostatistical literature, simulating a realisation of a spatial process $S(x)$ on a set of locations $x_i : i = 1, \ldots, n$ is called *unconditional simulation*, to distinguish it from *conditional simulation*. The latter refers to simulation of a spatial process $S(x)$ at locations $x_i^* : i = 1, \ldots, N$, conditional on observed values $S(x_i)$ at locations $x_i : i = 1, \ldots, n$ or, more generally, conditional on data $Y = (Y_1, \ldots, Y_n)$ which are stochastically related to $S(\cdot)$. In the present context, the underlying model for Y is that $Y_i = S(x_i) + Z_i$, where the Z_i are mutually independent and normally distributed, $Z_i \sim N(0, \tau^2)$. Conditional simulation is used informally to investigate to what extent the observed data do or do not identify the essential features of the underlying spatially continuous surface $S(x)$. It is also an essential tool in formal geostatistical inference, and as such will arise naturally in later chapters. Here, we note only that for the Gaussian model, the conditional distribution of the values of the process $S(x)$ at any set of locations, say $S^* = \{S(x_i^*), .., S(x_N^*)\}$, given the data Y, is multivariate Gaussian with a variance matrix which does not depend on Y. Hence, both unconditional and conditional simulation require computationally feasible ways of simulating from high-dimensional multivariate Gaussian distributions with particular kinds of structured covariance matrices.

3.11   Low-rank models

A low-rank model (Hastie, 1996) for a random vector S is one whose distributional dimension is less than the dimension of S itself. To motivate this idea in the context of geostatistical modelling, we briefly re-visit the singular value decomposition method for simulating realisations of S when the underlying model is a Gaussian process.

Recall that the singular value decomposition method simulates S as $S = AZ$ where Z is a vector of mutually independent $N(0, 1)$ random variables and $A = U\Lambda^{\frac{1}{2}}$. Here, the diagonal matrix Λ contains the eigenvalues of the required covariance matrix of S, whilst U contains the corresponding eigenvectors. If the eigenvalues are ordered from largest to smallest, then we could obtain an approximate simulation of S by using only the first $m < n$ columns of A to give

$$S = A_m Z \tag{3.22}$$

where now Z consists of only m independent $N(0, 1)$ variates (see Exercise 3.4). The resulting S has a singular multivariate Gaussian distribution, which can be regarded as a low-rank approximation to the target, non-singular distribution. Because A is derived from the covariance matrix of $S = \{S(x_1), \ldots, S(x_n)\}$ its elements are, implicitly, functions of the sampling locations x_i and we could therefore think of (3.22) as a specification of the form

$$S(x_i) = \sum_{j=1}^{m} Z_j f_j(x_i) : i = 1, \ldots, n. \tag{3.23}$$

This suggests that, rather than considering the low-rank approximation only as a computational short-cut, we could also use it as a way of defining a model for $S(\cdot)$. The general idea is to represent a spatially continuous stochastic process $S(x)$ as a linear combination of functions $f_j(x)$ and random coefficients A_j, so that for any $x \in \mathbb{R}^2$,

$$S(x) = \sum_{j=1}^{m} A_j f_j(x). \tag{3.24}$$

If the A_j follow a zero-mean multivariate Gaussian distribution with $\mathrm{Cov}(A_j, A_k) = \gamma_{jk}$, then $S(\cdot)$ is a zero-mean Gaussian process with covariance structure given by

$$\mathrm{Cov}\{S(x), S(x')\} = \sum_{j=1}^{m}\sum_{k=1}^{m}\gamma_{jk}f_j(x)f_k(x'). \tag{3.25}$$

In general, the covariance structure (3.25) is non-stationary. Whether or not it has an intuitively appealing form depends on the choices made for the functions $f_j(\cdot)$ and for the covariances amongst the A_j. The $f_k(\cdot)$ would usually be chosen to form an orthonormal basis, meaning that

$$\int f_j(x)f_k(x)dx = 1$$

if $k = j$ and is zero otherwise. Typically, the coefficients A_j would then be specified as mutually independent.

A familiar example of (3.24) in one dimension is the spectral representation of a time-series as a superposition of sine and cosine waves with mutually independent random coefficients. For an exact representation of a time-series $S(x) : x = 1, .., n$ we define n functions $f_k(x)$ which correspond to $n/2$ sine-cosine pairs at frequencies $2\pi jx/n : j = 0, 1, \ldots, [n/2]$. The associated coefficients are then assigned large or small variances corresponding to frequencies which account for large or small proportions, respectively, of the overall variation in the series. A low-rank approximation is obtained by setting some of the coefficients to zero. Spectral representations can also be used in two spatial dimensions and are discussed for example in Stein (1999).

Low-rank models for spatial processes can also be constructed using splines. Splines (Wahba, 1990) are piece-wise polynomial functions. By choosing the pieces to be cubics, constrained to be continuously differentiable at the joins, or "knots" connecting successive pieces, we obtain a very flexible method for approximating any smooth function. In two spatial dimensions, the same idea can be used to construct a flexible class of smooth surfaces by joining together locally polynomial pieces, known as *thin-plate splines* (Duchon, 1977). Thin-plate spline models are discussed in Wood (2003). Kammann and Wand (2003) emphasise the connection between splines and linear random effect models which is hinted at in (3.24) above. Laslett (1994) compares predictions obtained from spline models and from more conventional geostatistical models of the kind discussed earlier in this chapter. Ruppert, Wand and Carroll (2003) discuss the use of low-rank splines in semiparametric regression modelling.

3.12   Multivariate models

Multivariate geostatistical models are relevant when two or more different response variables are measured at spatial locations within a continuous spatial region. As discussed in Section 1.2.2 this situation can arise either because the variables are all of equal scientific interest and we wish to describe their joint spatial distribution, or because we wish to describe the conditional distribution of a response variable of primary interest given one or more spatially referenced covariates. When a covariate is only available at a finite set of sample locations we may choose to treat it as a set of sampled values from an underlying stochastic process. A third situation in which multivariate methods are useful is when the variable of primary interest, Y say, is difficult or expensive to measure, but it is easy to measure a second variable, Z, which is known to be correlated with Y. In this situation, for efficient prediction of Y the most cost-effective design may be one in which a small number of measurements of Y are combined with a large number of cheaper measurements of Z.

In the remainder of this section we describe some possible multivariate extensions to the univariate Gaussian models considered so far in this chapter. All of the general ideas discussed for univariate processes carry over, but with additional aspects introduced by the multivariate setting. We focus on the specification of valid models for stationary variation about a trend, including the distinction between the observation process $Y(x)$ and an unobserved signal process $S(x)$.

3.12.1   Cross-covariance, cross-correlation and cross-variogram

The covariance and correlation functions of a multivariate spatial process are easily defined as follows. A d-dimensional spatial process is a collection of random variables $Y(x) = \{Y_1(x), \ldots, Y_d(x)\}$, where $x \in \mathbb{R}^2$. Then, the covariance function of $Y(x)$ is a $d \times d$ matrix-valued function $\Gamma(x, x')$, whose $(j, k)^{th}$ element is

$$\gamma_{jk}(x, x') = \text{Cov}\{Y_j(x), Y_k(x')\}. \tag{3.26}$$

For each pair of locations (x, x'), the matrix $\Gamma(x, x')$ is symmetric i.e., $\gamma_{jk}(x, x') = \gamma_{kj}(x, x')$.

When $Y(x)$ is stationary, $\gamma_{jj}(x, x) = \text{Var}\{Y_j(x)\} = \sigma_j^2$ does not depend on x, and for $j \neq k$, $\gamma_{jk}(x, x')$ depends only on $u = ||x - x'||$. We then define the correlation function of $Y(x)$ as the matrix-valued function $R(u)$ whose $(j, k)^{th}$ element is $\rho_{jk}(u) = \gamma_{jk}(u)/(\sigma_j \sigma_k)$. When $k = j$, the functions $\rho_{jj}(u)$ are the correlation functions of the univariate processes $Y_j(x)$ and are symmetric in u i.e., $\rho_{jj}(-u) = \rho_{jj}(u)$. When $k \neq j$, the functions $\rho_{jk}(u)$, called the cross-correlation functions of $Y(x)$, are not necessarily symmetric but must satisfy the condition that $\rho_{jk}(u) = \rho_{kj}(-u)$.

To define a cross-variogram for $Y(x)$, there are at least two possibilities. The first, and the more traditional, is

$$V_{jk}^*(u) = \frac{1}{2}\text{Cov}[\{Y_j(x) - Y_j(x - u)\}\{Y_k(x) - Y_k(x - u)\}]. \tag{3.27}$$

See, for example, Journel and Huijbregts (1978) or Chilès and Delfiner (1999). Expanding the right-hand side of (3.27) we find that

$$
\begin{aligned}
V_{jk}^*(u) &= \gamma_{jk}(0) - \frac{1}{2}\{\gamma_{jk}(u) + \gamma_{jk}(-u)\} \\
&= \sigma_j\sigma_k[1 - \frac{1}{2}\{\rho_{jk}(u) + \rho_{jk}(-u)\}].
\end{aligned}
\tag{3.28}
$$

The similarity between (3.28) and the corresponding relationship between univariate covariance, correlation and variogram functions, as discussed in Section 3.4, is clear.

The second possibility, introduced by Cressie and Wikle (1998) and called by them the *variance-based cross-variogram*, is

$$
V_{jk}(u) = \frac{1}{2}\text{Var}\{Y_j(x) - Y_k(x - u)\}.
\tag{3.29}
$$

Expanding the right-hand side of (3.29) gives

$$
V_{jk}(u) = \frac{1}{2}(\sigma_j^2 + \sigma_k^2) - \sigma_j\sigma_k\rho_{jk}(u).
\tag{3.30}
$$

The expansion (3.30) highlights an apparent objection to (3.29), namely that it mixes incompatible physical dimensions. However, we can overcome this by working with standardised, and therefore dimensionless, variables. An advantage of (3.29) over (3.27) is that it suggests a way of estimating the variogram empirically which does not require the different variables to be measured at a common set of sampling locations.

Using standardised variables reduces the two definitions of the cross-variogram in (3.30) and (3.28) to

$$
V_{jk}^*(u) = 1 - \frac{1}{2}\{\rho_{jk}(u) + \rho_{jk}(-u)\}
$$

and

$$
V_{jk}(u) = 1 - \rho_{jk}(u),
$$

respectively, hence

$$
V_{jk}^*(u) = \frac{1}{2}\{V_{jk}(u) + V_{jk}(-u)\}.
$$

In particular, provided that we use standardised variables, we see that $V_{jk}^*(u) = V_{jk}(u)$ whenever the cross-correlation function $\rho_{jk}(u)$ is symmetric in u.

3.12.2 Bivariate signal and noise

To construct a stationary Gaussian model for bivariate data $(Y_{ij} : i = 1, ..., n_j, j = 1, 2)$ measured at locations x_{ij} we first specify a model for an unobserved bivariate stationary Gaussian process $\{S(x) = (S_1(x), S_2(x)) : x \in \mathbb{R}^2\}$, with bivariate mean zero, variances $\sigma_j^2 = \text{Var}\{S_j(x)\}$ and correlation structure determined by three functions $\rho_{11}(u) = \text{Corr}\{S_1(x), S_1(x - u)\}$, $\rho_{22}(u) = \text{Corr}\{S_2(x), S_2(x - u)\}$ and $\rho_{12}(u) = \text{Corr}\{S_1(x), S_2(x - u)\}$.

The simplest assumption we can make about the data Y_{ij} is that $Y_{ij} = S_j(x_{ij})$ i.e., the signal at any location x can be observed without error. When the data are subject to measurement error, the simplest assumption is that the Y_{ij} are mutually independent given $S(\cdot)$ and normally distributed,

$$Y_{ij} \sim \mathrm{N}\{\mu_j(x_{ij}) + S_j(x_{ij}), \tau_j^2\} : i = 1, \ldots, n_j; j = 1, 2. \qquad (3.31)$$

Under this model, each dimension of the response separately follows a univariate Gaussian model, whilst dependence between the two response dimensions is modelled indirectly through the structure of the unobserved process $S(\cdot)$. The conditional independence assumption in (3.31) invites the interpretation that the parameters τ_j^2 represent the measurement error variances in each of the two response dimensions. A less restrictive assumption than (3.31) would be to allow the measurement errors associated with $Y(x) = \{Y_1(x), Y_2(x)\}$ to be correlated. This would only affect the model at locations where both of $Y_1(x)$ and $Y_2(x)$ are measured; where only one of the $Y_j(x)$ is measured, (3.31) would still hold. Correlated measurement errors might be particularly appropriate if, as already discussed in the univariate setting, we want the nugget effect to include spatial variation on scales smaller than the smallest inter-point distance in the sampling design.

In the case of spatially independent error terms, the mean and covariance structure of the data, Y_{ij}, are given by

$$\mathrm{E}[Y_{ij}] = \mu_j(x_{ij}),$$

$$\mathrm{Var}\{Y_{ij}\} = \tau_j^2 + \sigma_j^2$$

and, for $(i, j) \neq (i', j')$,

$$\mathrm{Cov}\{Y_{ij}, Y_{i'j'}\} = \sigma_j \sigma_{j'} \rho_{jj'}(\|x_{ij} - x_{i'j'}\|).$$

Note in particular that non-zero error variances τ_j^2 induce discontinuities at the origin in the covariance structure of the measurement process.

3.12.3 Some simple constructions

In order to construct particular bivariate models, we need to specify explicit forms for the two mean functions $\mu_j(x)$ and for the covariance structure of $S(\cdot)$. With regard to the means, in practice the easiest models to handle are those in which the means are linear functions of spatial explanatory variables, as was also true in the univariate case. With regard to the covariance structure, the univariate models discussed earlier are a natural starting point. However, in extending these to the bivariate case, we need to be sure that the required positive definiteness conditions are not violated. Note that these require that arbitrary linear combinations of either or both of the response dimensions should have non-negative variances. A simple way to ensure that this is the case is to build a bivariate model explicitly from univariate components. The same holds, with the obvious modifications, for multivariate processes of dimension $d > 2$.

A common-component model

One example of an explicit bivariate construction is the following. Suppose that $S_0^*(\cdot)$, $S_1^*(\cdot)$ and $S_2^*(\cdot)$ are independent univariate stationary Gaussian processes with respective covariance functions $\gamma_j(u) : j = 0, 1, 2$. Define a bivariate process $S(\cdot) = \{S_1(\cdot), S_2(\cdot)\}$ to have components

$$S_j(x) = S_0^*(x) + S_j^*(x) : j = 1, 2.$$

Then, by construction, $S(\cdot)$ is a valid bivariate process with covariance structure

$$\text{Cov}\{S_j(x), S_{j'}(x - u)\} = \gamma_0(u) + I(j = j')\gamma_j(u)$$

where $I(\cdot)$ is the indicator function, equal to one if its logical argument is true, zero otherwise. Note that if, as is typically the case, the covariance functions $\gamma_j(u)$ are non-negative valued, then this construction can only generate non-negative cross-covariances between $S_1(\cdot)$ and $S_2(\cdot)$. In practice this is often the case or, if the two variables are inversely related, can be made so by reversing the sign of one of the components. The common-component construction extends to processes of dimension $d > 2$ in which all of the components $S_j(x)$ share an underlying common component $S_0^*(x)$. Note, however, that the simple device of applying a change of sign to $S_0(x)$ obviously cannot induce an arbitrary mix of positive and negative cross-covariances. Also, as written the construction implicitly assumes a common measurement scale for all of the component processes. When this is not the case, the model requires an additional $d - 1$ scaling parameters so that the common component $S_0^*(x)$ is replaced by $S_{0j}^*(x) = \sigma_{0j} R(x)$ where $R(x)$ has unit variance.

Linear combinations of independent components

Another simple construction is to begin with two, or more generally d, independent univariate processes $U_k(x)$ and define $S_j(x)$ as a linear combination,

$$S_j(x) = \sum_{j=1}^{d} a_{kj} U_j(x),$$

or in vector-matrix notation,

$$S(x) = AU(x). \tag{3.32}$$

Without loss of generality, we can assume that each process $U_k(x)$ has unit variance. If $U_k(x)$ has correlation function $\rho_k(\cdot)$, it follows that the matrix-valued covariance function of $S(x)$ is

$$\Gamma(x, x') = ARA', \tag{3.33}$$

where R is the diagonal matrix with diagonal entries $R_{kk} = \rho_k(x - x')$. In the special case where $\rho_k(u) = \rho(u)$, (3.33) reduces to $\Gamma(x, x') = B\rho(x-x')$. This is sometimes called the *proportional covariance model* (Chilès and Delfiner, 1999). The assumption that all of the $U_k(x)$ share a common correlation function reduces the number of parameters in the model to manageable proportions, but otherwise does not seem particularly natural.

Schmidt and Gelfand (2003) use a variant of (3.32) in which there is a natural ordering of the components of $S(x)$ so that $S_1(x)$ depends on $U_1(x)$ only, $S_2(x)$ depends on $U_1(x)$ and $U_2(x)$, and so on. Gelfand, Schmidt, Banerjee and Sirmans (2004) extend this model to allow the non-zero elements of the A_i to depend on location, x.

The linear model of co-regionalisation

By construction, we can also obtain valid models by adding linear combinations of $p \geq 2$ models with independent components. Hence, we can define a model for a d-dimensional process $S(x)$ as

$$S(x) = \sum_{i=1}^{p} A_i U^i(x), \tag{3.34}$$

where now each $U^i(x) = \{U_1^i(x), \ldots, U_d^i(x)\}$ is a set of d independent univariate processes and A_i is a $d \times d$ matrix. In practice, models of this kind would be very poorly identified without some restrictions being placed beforehand on the processes $U_k^i(x)$. In the *linear model of co-regionalisation*, these restrictions are that each term on the right-hand side of (3.34) is a proportional covariance model. This again raised the question of whether the resulting savings in the number of unknown parameters has a natural scientific interpretation or is merely a pragmatic device.

How useful are standard classes of multivariate model?

The question is worth asking because, as the examples above illustrate, even very simple multivariate constructions quickly lead to models with either large numbers of parameters and consequent problems of poor identifiability, or potentially severe restrictions on the allowable form of cross-correlation structure. A better modelling strategy than an empirical search through a richly parameterised standard model class may be to build multivariate models by incorporating structural assumptions suggested by the context of each specific application; see, for example, Knorr-Held and Best (2001), who use the common component model in an epidemiological setting where it has a natural interpretation.

3.13   Computation

We first show how to use **geoR** to compute and plot standard correlation functions. The function `cov.spatial()` has an argument `cov.model` which allows the user to choose from a set of correlation families. Options include the Matérn, powered exponential, spherical and wave families discussed earlier in this chapter; a complete list can be obtained by typing *help(cov.spatial)*. Below, we show the commands used to produce Figure 3.2. Similar commands were used for Figure 3.5 and Figure 3.7.

```
> x <- seq(0, 1, l = 101)
> plot(x, cov.spatial(x, cov.model = "mat", kappa = 0.5,
+       cov.pars = c(1, 0.25)), type = "l", xlabel = "u",
+       ylabel = expression(rho(u)), ylim = c(0, 1))
> lines(x, cov.spatial(x, cov.model = "mat", kappa = 1.5,
+       cov.pars = c(1, 0.16)), lty = 2)
> lines(x, cov.spatial(x, cov.model = "mat", kappa = 2.5,
+       cov.pars = c(1, 0.13)), lty = 3)
```

We now illustrate the use of the **geoR** function grf() for generating simulations of two-dimensional Gaussian processes. We encourage the reader to experiment with different input parameters so as to obtain an intuitive understanding of the different ways in which the model parameters affect the appearance of the simulated realisations. The arguments to grf() specify the model and the locations for which simulated values are required. The locations can be specified to form a regular lattice, a completely random pattern, or a configuration supplied explicitly as a set of (x, y) coordinates. For example, to produce Figure 3.4 we used the following commands.

```
> set.seed(159)
> image(grf(100^2, grid = "reg", cov.pars = c(1, 0.25)),
+       col = gray(seq(1, 0, l = 51)), xlab = "", ylab = "")
> set.seed(159)
> image(grf(100^2, grid = "reg", cov.pars = c(1, 0.13),
+       cov.model = "mat", kappa = 2.5), col = gray(seq(1,
+       0, l = 51)), xlab = "", ylab = "")
```

Using the R function set.seed() ensures that simulations are generated with the same random number seed, hence differences between the simulated realisations are due only to the different values of the model parameters. In the example above, the realisation covers $n = 100^2 = 10,000$ locations, whilst the argument grid="reg" instructs the function to generate the locations in a 100 by 100 regular square lattice.

For the simulations of the anisotropic model in Figure 3.10 we used the argument aniso.pars to specify the anisotropy angle and ratio, as follows.

```
> set.seed(421)
> image(grf(201^2, grid = "reg", cov.pars = c(1, 0.25),
+       aniso.pars = c(pi/3, 4)), col = gray(seq(1, 0, l = 51)),
+       xlab = "", ylab = "")
> set.seed(421)
> image(grf(201^2, grid = "reg", cov.pars = c(1, 0.25),
+       aniso.pars = c(3 * pi/4, 2)), col = gray(seq(1, 0,
+       l = 51)), xlab = "", ylab = "")
```

The function grf() allows the user to select from several algorithms for generating the simulated realisations, including an automatic link to the function GaussRF() within the R package **RandomFields** written by Martin Schlather. To invoke this link, the user specifies the optional argument method="RF" in

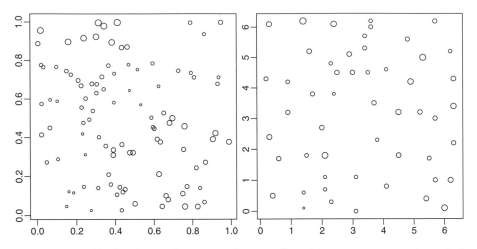

Figure 3.16. Realisations of two stationary Gaussian processes on irregularly distributed sets of locations. See text for detailed specifications.

the call to the function grf(). At the time of writing, the default in the latest version of the **geoR** package is to use the Choleski factorisation for $n \leq 500$, and the link to GaussRF() for $n > 500$. The **RandomFields** package is also available at the CRAN website, *http://cran.r-project.org*.

Note also that Håvard Rue has written very efficient code, available for download at *http://www.math.ntnu.no/~hrue/GMRFLib*, for simulation of Gaussian processes on very large numbers of locations using an approximating Markov random field, as described in Section 3.10. Rue and Held (2005) provide details on the methods and on the use of the software.

Figure 3.16 shows two further examples of simulations generated by grf(), using the commands below. The first call to the function produces the simulation shown in the left-hand panel, a realisation of a stationary Gaussian model with mean $\mu = 0$, variance $\sigma^2 = 1$ and Matérn correlation function with $\kappa = 1.5$ and $\phi = 0.15$. The simulation generates 100 values at locations distributed completely at random over the unit square. The right panel shows simulated values at the 52 locations of the elevation data from Example 1.1. In this case, we have used a stationary Gaussian model with mean $\mu = 850$, nugget variance $\tau^2 = 100$, signal variance $\sigma^2 = 3500$ and Matérn correlation function with $\kappa = 2.5$ and $\phi = 0.8$.

```
> sim1 <- grf(100, cov.pars = c(1, 0.15), cov.model = "matern",
+       kappa = 1.5)
> points(sim1)
> data(elevation)
> sim2 <- grf(grid = elevation$coords, cov.pars = c(3500,
+       0.8), nugget = 100)
> sim2$data <- sim2$data + 850
> points(sim2)
```

3.14   Exercises

3.1. Consider a one-dimensional spatial process $S(x) : x \in \mathbb{R}$ with mean μ, variance σ^2 and correlation function $\rho(u) = \exp(-u/\phi)$. Define a new process $R(x) : x \in \mathbb{R}$ by the equation

$$R(x) = (2\theta)^{-1} \int_{x-\theta}^{x+\theta} S(u)du.$$

Derive the mean, variance and correlation function of $R(\cdot)$. Comment briefly.

3.2. Is the following a legitimate correlation function for a one-dimensional spatial process $S(x) : x \in \mathbb{R}$?

$$\rho(u) = \begin{cases} 1 - u & : \quad 0 \le u \le 1 \\ 0 & : \quad u > 1 \end{cases}$$

Give either a proof or a counter-example.

3.3. Derive a formula for the volume of the intersection of two spheres of equal radius, ϕ, whose centres are a distance u apart. Compare the result with the formula (3.8) for the spherical variogram and comment.

3.4. Consider the following method of simulating a realisation of a one-dimensional spatial process on $S(x) : x \in \mathbb{R}$, with mean zero, variance 1 and correlation function $\rho(u)$. Choose a set of points $x_i \in \mathbb{R} : i = 1, \ldots, n$. Let R denote the correlation matrix of $S = \{S(x_1), \ldots, S(x_n)\}$. Obtain the singular value decomposition of R as $R = D\Lambda D'$ where λ is a diagonal matrix whose non-zero entries are the eigenvalues of R, in order from largest to smallest. Let $Y = \{Y_1, \ldots, Y_n\}$ be an independent random sample from the standard Gaussian distribution, $N(0, 1)$. Then the simulated realisation is

$$S = D\Lambda^{\frac{1}{2}}Y. \tag{3.35}$$

Write an R function to simulate realisations using the above method for any specified set of points x_i and a range of correlation functions of your choice. Use your function to simulate a realisation of S on (a discrete approximation to) the unit interval $(0, 1)$.

Now investigate how the appearance of your realisation S changes if in (3.35) you replace the diagonal matrix Λ by a truncated form in which you replace the last k eigenvalues by zeros.

3.5. Consider a spatial process $S(\cdot)$ defined by

$$S(x) = \int w(u)S^*(x - u)du$$

where $w(u) = (2\pi)^{-1} \exp(-||u||^2/2)$ and $S^*(\cdot)$ is another stationary Gaussian process. Derive an expression for the correlation function, $\rho(u)$ say, of $S(\cdot)$ in terms of $w(\cdot)$ and the correlation function, $\rho^*(u)$ say, of $S^*(\cdot)$. Give explicit expressions for $\rho(u)$ when $\rho^*(u)$ is of the form:

(a) pure nugget, $\rho^*(u) = 1$ if $u = 0$, zero otherwise;
(b) spherical;
(c) Gaussian.
(d) In each case, comment on the mean square continuity and differentiability properties of the process $S(\cdot)$ in relation to its corresponding $S^*(\cdot)$.

4

Generalized linear models for geostatistical data

4.1 General formulation

In the classical setting of independently replicated data, the generalized linear model (GLM) as introduced by Nelder and Wedderburn (1972) provides a unifying framework for regression modelling of continuous or discrete data. The original formulation has since been extended, in various ways, to accommodate dependent data. In this chapter we enlarge on the brief discussion of Section 1.4 to consider extensions of the classical GLM which are suitable for geostatistical applications.

The basic ingredients of a GLM are the following:

1. *responses* $Y_i : i = 1, \ldots, n$ are mutually independent with expectations μ_i;

2. the μ_i are specified by $h(\mu_i) = \eta_i$, where $h(\cdot)$ is a known *link function* and η_i is a *linear predictor*, $\eta_i = d'_i\beta$; in this last expression, d_i is a vector of explanatory variables associated with the response Y_i and β is a vector of unknown parameters;

3. the Y_i follow a common distributional family, indexed by their expectations, μ_i, and possibly by additional parameters common to all n responses.

Working within this framework, Nelder and Wedderburn (1972) showed how a single algorithm could be used for likelihood-based inference. This enabled the development of a single software package, GLIM, for fitting any model within the GLM class. The fitting algorithm was subsequently incorporated into many general-purpose statistical packages, including the glm() function within R. GLM's occupy a central place in modern applied statistics.

One of a number of ways to extend the GLM to accommodate dependent responses is to introduce unobservable *random effects* into the linear predictor. Thus, in the second part of the model specification above, η_i is modified to

$$\eta_i = d'_i\beta + S_i$$

where now $S = (S_1, \ldots, S_n)$ follows a zero-mean multivariate distribution. The S_i are called *random effects* or *latent variables*. Models of this kind are called generalized linear mixed models (GLMM's). Breslow and Clayton (1993) give further details and a range of applications. In practice, the most common specification for S is as a multivariate Gaussian random variable with a particular covariance structure imposed according to the practical context.

In a GLMM, the simplest assumption we could make about the S_i is that they are mutually independent, in which case the model is sometimes said to incorporate extra-variation, or over-dispersion, relative to the corresponding classical GLM. For example, when a Poisson log-linear model is fitted to independent count data, it is often found that in an otherwise well-fitting model the variance is larger than the mean, whereas the Poisson assumption implies that they should be equal. A GLMM with mutually independent S_i is one of several ways to account for this effect, which is often called *extra-variation* or *over-dispersion*. To model dependent data using a GLMM, we need to specify a suitable form of dependence amongst the S_i. For example, in longitudinal studies where the Y_i arise as repeated measurements taken from many different individuals, it is usual to assume that the S_i are independent between individuals but correlated within individuals. The statistical methods associated with models of this kind can exploit the independent replication between individuals in order to check directly any assumed form for the correlation structure within subjects, or to develop methods of analysis which are in some respects robust to mis-specification of the correlation structure. See, for example, Diggle, Heagerty, Liang and Zeger (2002), in particular their discussion of *marginal models* for longitudinal data.

For geostatistical applications, we usually cannot rely on any form of independent replication. Instead, the observed responses $y = (y_1, \ldots, y_n)$ must be considered as a single realisation of an n-dimensional random variable Y. In this setting, we shall use GLMM's in which S equates to $S = \{S(x_1), \ldots, S(x_n)\}$, the values of an underlying Gaussian signal process at each of the sample locations x_i. This very natural extension of GLMM's was investigated systematically by Diggle et al. (1998). We shall refer to a model of this kind as a generalized linear geostatistical model, or GLGM. This is not the only way in which we could adapt the classical GLM for use in geostatistical applications, but it is the approach on which we shall focus most of our attention.

The generalized linear modelling strategy is most appealing when the distributional family for the responses Y_i, conditional on the random effects S in the case of a mixed model, follows naturally from the sampling mechanism. For this reason, two of the most widely used GLM's are the Poisson log-linear model for count responses, and the logistic-linear model for binary, or more generally binomial, responses. For geostatistical applications, the same philosophy applies. In particular, we advocate the use of GLGM's only as a way of incorporating explicit knowledge of the sampling mechanism which generates the data. When the need is to address empirical departure from linear Gaussian assumptions, for example when continuous-valued measurement data exhibit a strongly skewed distribution, our preferred initial modelling framework would be the transformed Gaussian model as discussed in Chapter 3.

In the remainder of this chapter, we first consider the form of the theoretical variogram for a stationary GLGM. This gives some insight into the statistical properties of this class of models, but can also be helpful for exploratory data analysis using the empirical variogram. We then describe the two most widely used examples of GLGM's, namely the Poisson log-linear and the binomial logistic-linear, followed by a short discussion of spatial models for survival data. We describe some of the connections between GLGM's and spatial point process models, including the log-Gaussian Cox Process (Møller, Syversveen and Waagepetersen, 1998) and a possible approach to dealing with preferentially sampled geostatistical data. We end the chapter with some examples of spatially continuous models which fall outside the GLGM class.

4.2 The approximate covariance function and variogram

The variogram is based on second-order moments, and therefore gives a very natural way to describe the dependence structure in a Gaussian model. In non-Gaussian settings, the variogram is a less natural summary statistic but can still be useful as a diagnostic tool. The approximate form of the variogram for a non-Gaussian GLGM is therefore of some interest. Here, we consider only the stationary form of the model, in which there are no spatial trends.

We suppose that $S(x)$ is a stationary Gaussian process with mean zero and variance σ^2, and that the observations Y_i, conditional on $S(\cdot)$, are mutually independent with conditional expectations $\mu_i = g(\alpha + S_i)$ and conditional variances $v_i = v(\mu_i)$. Here, S_i is shorthand notation for $S(x_i)$ and $g(\cdot)$ is the analytic inverse of the link function, $h(\cdot)$. Then, the Y-variogram is $\gamma_Y(u) = \mathrm{E}[\frac{1}{2}(Y_i - Y_j)^2]$, where $u = ||x_i - x_j||$. Using standard conditional expectation arguments, we have that

$$
\begin{aligned}
\gamma_Y(u) &= \frac{1}{2}\mathrm{E}_S[\mathrm{E}_Y[(Y_i - Y_j)^2|S(\cdot)]] \\
&= \frac{1}{2}\mathrm{E}_S[\{g(\alpha + S_i) - g(\alpha + S_j)\}^2 + v(g(\alpha + S_i)) + v(g(\alpha + S_j))] \\
&= \frac{1}{2}\left(\mathrm{E}_S[\{g(\alpha + S_i) - g(\alpha + S_j)\}^2] + 2\mathrm{E}_S[v(g(\alpha + S_i))]\right), \quad (4.1)
\end{aligned}
$$

where the last equality follows because the marginal distribution of $S(x_i)$ is the same for all locations x_i. The second term on the right-hand side of (4.1) is a constant, which we write as $2\bar{\tau}^2$. This choice of notation emphasises that $\bar{\tau}^2$, obtained by averaging a conditional variance over the distribution of $S(\cdot)$, is analogous to the nugget variance in the stationary Gaussian model. To approximate the first term on the right-hand-side, we use a first-order Taylor series approximation $g(\alpha + S) \approx g(\alpha) + Sg'(\alpha)$, to give the result

$$\gamma_Y(u) \approx g'(\alpha)^2 \gamma_S(u) + \bar{\tau}^2. \tag{4.2}$$

In other words, the variogram on the Y-scale is approximately proportional to the variogram of the latent Gaussian process $S(\cdot)$, plus an intercept which represents an average nugget effect induced by the variance of the error distribution of the model.

Note that (4.2) relies on a linear approximation to the inverse link function, $g(\cdot)$. Although this leads to a helpful interpretation in terms of the effective nugget variance, $\bar{\tau}^2$, it may be inadequate for diagnostic analysis since the essence of the generalized linear model family is its explicit incorporation of a non-linear relationship between Y and $S(x)$. The exact variogram on the Y-scale necessarily depends on higher moments of the latent process $S(\cdot)$. As we shall see in later chapters, explicit results are available in special cases.

4.3   Examples of generalised linear geostatistical models

4.3.1   The Poisson log-linear model

The Poisson log-linear model, as its name implies, is a GLM in which the link function is the logarithm and the conditional distribution of each Y_i is Poisson. The model is a natural candidate for spatially referenced count data like the Rongelap data of Example 1.2, where the local mean of a Poisson count is determined by the value of an unobserved, real-valued stochastic process; in the Rongelap example, the unobserved process represents a spatially varying level of residual contamination. In the simplest form of the model, the Y_i are conditionally independent Poisson counts with conditional expectations μ_i, where

$$\log \mu_i = \alpha + S(x_i) \tag{4.3}$$

and $S(\cdot)$ is a stationary Gaussian process with mean zero, variance σ^2 and correlation function $\rho(u)$.

Figure 4.1 shows a simulation in which the data are observed at 2500 locations in a 50 by 50 grid. In each of the two cases shown, the contours represent the conditional expectation surface, $\exp\{\alpha + S(x)\}$, whilst the grey-scale corresponds to the Poisson count, Y_i, associated with the location at the centre of each grid-square. The realisation of $S(\cdot)$ is the same in the two cases, and is generated from a Gaussian process with zero mean, variance $\sigma^2 = 2$ and Matérn correlation function with parameters $\kappa = 1.5$ and $\phi = 0.18$. In the left-hand panel $\alpha = 0.5$ whereas in the right-hand panel $\alpha = 5$. When α is small, the Poisson variation dominates the signal and the grey-scale piece-wise constant surface based on the counts bears only a mild resemblance to the contour representation of the underlying conditional expectation surface. In contrast, when α is large the grey-scale and contour surfaces are in closer correspondence.

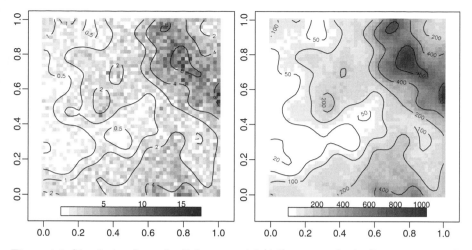

Figure 4.1. Simulating from the Poisson model (4.3); grey-scale shading represents the data values on a regular grid of sampling locations, whilst the contours represents the conditional expectation surface, with $\alpha = 0.5$ on the left panel and $\alpha = 5$ on the right panel.

Note that in the Poisson model, unlike the linear Gaussian model, the conditional variance of Y_i given $S(x_i)$ is not a free parameter, but is constrained to be equal to the conditional expectation of Y_i. In practice, we may well encounter evidence of additional variability in the data, often called extra-Poisson variation, which is not spatially structured. In this case, a natural extension to the model is to include a nugget effect within the linear predictor. The conditional distribution of the Y_i is then still modelled as Poisson with conditional expectations μ_i, but (4.3) is extended to

$$\log \mu_i = \alpha + S(x_i) + Z_i \tag{4.4}$$

where $S(\cdot)$ is as before and the Z_i are mutually independent $N(0, \tau^2)$. In principle, this extension of the model allows us to disentangle two components of the nugget variance which were generally indistinguishable in the linear Gaussian model: the Poisson variation induced by the sampling scheme, analogous to our earlier interpretation of the nugget effect as measurement error, and a spatially uncorrelated component analogous to the alternative interpretation of the nugget effect as small-scale spatial variation.

4.3.2   The binomial logistic-linear model

In this model, the link function is the logit, and the responses Y_i represent the outcomes of conditionally independent Bernoulli trials with $P\{Y_i = 1 | S(\cdot)\} = p(x_i)$, where, in the stationary case,

$$\log[p(x_i)/\{1 - p(x_i)\}] = \alpha + S(x_i).$$

The information content in data generated from this model is rather limited unless the intensity of the sample locations is large relative to the variation

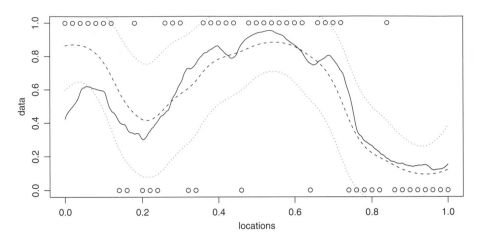

Figure 4.2. Simulation of a binary-valued logistic-linear model. The solid line shows the function $p(x) = \exp\{S(x)\}/[1 + \exp\{S(x)\}]$, where $S(\cdot)$ is a stationary Gaussian process. The open circles are the realised values of a binary sequence Y_i with $\mathrm{P}(Y_i = 1) = p(x_i)$ and x_i equally spaced over the unit interval. The dashed line shows predicted values using true model parameters, whilst the dotted lines show the corresponding pointwise 95% prediction intervals.

in the signal process $S(\cdot)$. For example, Figure 4.2 shows a one-dimensional simulation with binary responses Y_i obtained at 51 locations equally spaced along the unit interval. The intercept parameter in the linear predictor is $\alpha = 0$, and $S(x)$ is a Gaussian process with mean zero, variance $\sigma^2 = 5$ and Matérn correlation function with $\kappa = 1.5$ and $\phi = 0.1$.

In this example, we cannot expect to obtain a useful prediction of the continuous trace of $S(x)$ using only the observed values of the binary sequence Y_i. This is confirmed by the information displayed in Figure 4.2. The minimum mean square error predictor of $p(x)$, indicated in Figure 4.2 by a dashed line, shows only modest deviations from the *a priori* mean of $p(x)$ relative to the width of the pointwise 95% prediction intervals.

In practice, in a geostatistical setting the binomial model is much more useful when the binary Y_i are replaced by conditionally binomial counts with large denominators n_i. An example of this is provided by the Gambia malaria data of Example 1.3.

As with the Poisson log-linear model, we can also extend the model to incorporate a spatially uncorrelated extra-binomial variance component by adding a term Z_i to the linear predictor for Y_i such that the Z_i are mutually independent $N(0, \tau^2)$.

4.3.3 Spatial survival analysis

Survival analysis is a very well established area of statistical methodology in its own right. As the name implies, it is widely used in medical applications to model the survival prognosis of patients with a potentially fatal medical condition. The core problem of survival analysis is to build and fit regression models for time-to-event outcomes in the presence of *censoring*. The most common form of censoring is right-censoring, when some time-to-event outcomes are not observed, but are known only to be greater than an observed censoring time. For example, survival studies will usually end before all of the patients have died. Typically, survival analysis models are specified through their hazard function, $h(t)$, whose intuitive interpretation is that $h(t)\delta t$ is the conditional probability that a patient will die in the interval $(t, t + \delta t)$, given that they have survived until time t. More formally, $h(t) = f(t)/\{1 - F(t)\}$ where $f(\cdot)$ and $F(\cdot)$ denote the probability density and cumulative distribution function of survival time.

By far the most widely used approach to modelling $h(t)$, at least in medical applications, is to use a semi-parametric formulation introduced by Cox (1972). In this approach, the hazard for the ith patient is modelled as

$$h_i(t) = \lambda_0(t) \exp(z_i'\beta) \tag{4.5}$$

where z_i is a vector of explanatory variables for patient i and $\lambda_0(t)$ is an unspecified baseline hazard function. This is known as a *proportional hazards model*, because for any two patients i and j, $h_i(t)/h_j(t)$ does not change over time.

Fully parametric models have also been suggested, in which case families of distributions which could be used to model the survival time include the gamma, Weibull and log-Gaussian. Note that (4.5) reduces to an exponential distribution i.e., a special case of the gamma, if $\lambda_0(t) = \lambda$, a constant.

Another key idea in survival analysis is *frailty*. This corresponds exactly to the more widely used term *random effects*, whereby the variation in survival times between individual patients with identical values for the explanatory variables is greater than can be explained by the assumed distributional model. From the perspective of generalized linear modelling, the most obvious way to incorporate frailty would be to introduce an unobserved random variable within the exponential in (4.5), so defining a conditional hazard model,

$$h_i(t) = \lambda_0(t) \exp(z_i'\beta + U_i), \tag{4.6}$$

where the random effects U_i are an independent random sample from a distribution, for example the Gaussian. Within survival analysis, it is conventional to express frailty as a multiplicative effect on the hazard, hence (4.6) would be re-expressed as $h_i(t) = \lambda_0(t)W_i \exp(z_i'\beta)$, and the W_i are called the *frailties* for the patients in the study. The Gaussian assumption for the U_i in (4.6) therefore corresponds, in the terminology of survival analysis, to a *log-Gaussian frailty model*. A more popular choice is a *gamma frailty model*, in which the W_i follow a gamma distribution. It is hard to think of any compelling scientific reason for preferring the gamma to the log-Gaussian, or *vice versa*. A pragmatic reason for preferring the gamma is that it allows a closed form expression for the unconditional hazard function. Whatever distribution is assumed, frailties are scaled so that their expectation is one. Book-length accounts of models for survival data include Cox and Oakes (1984) and Hougaard (2000).

In the context of this chapter, the natural way to incorporate spatial effects into a hazard model is to replace the independent random sample $U_1, \ldots, U_n$ in (4.6) by a sample from an unobserved Gaussian process, hence

$\{S(x_1), \ldots, S(x_n)\}$, where x_i denotes the location of the ith patient. To pre-serve the interpretation of $\exp\{S(x)\}$ as a frailty process, we require $E[S(x)] = -0.5\text{Var}\{S(x)\}$. This is essentially the approach taken by Li and Ryan (2002) and by Banerjee, Wall and Carlin (2003). Li and Ryan (2002) preserve the semi-parametric setting of (4.6) and propose a "marginal rank likelihood" for making inferences about the regression parameters β which do not depend on the form of the baseline hazard. Banerjee et al. (2003) use a parametric model for the baseline hazard, in conjunction with Bayesian inference. This approach is closer in spirit to the general theme of this chapter, in which prediction of the unobserved, spatially varying frailties is assumed to be at least as important as inference about regression parameters. Both of these papers assume that the spatial resolution of the data is limited to regional level, hence they observe (possibly censored) survival time outcomes for a number of subjects in each of a discrete set of spatial regions which partition the study area. Henderson, Shi-makura and Gorst (2002) take a somewhat different approach. They consider how the widely used gamma-frailty model for independent survival outcomes can be modified to take account of spatial variation. In the case of regional-level spatial resolution, their individual-level frailties are conditionally independent and gamma-distributed within regions, given a set of regional mean frailties which are drawn from a multivariate Gaussian distribution with a spatially structured correlation matrix. They also suggest a way of generating spatially structured, gamma-distributed frailties at the individual level, albeit with some restrictions on the admissible parameters of the gamma marginal distributions, using the following construction. They assume that $Z_1, \ldots, Z_m$ are independent and identically distributed multivariate Gaussian random variables with mean zero, variance the identity matrix and spatially structured correlation matrix, $C = [c_{i,i'}]$. Then, writing each vector Z_j as $Z_j = (Z_{1j}, \ldots, Z_{nj})$ they define $W_i = \sum_{j=1}^m Z_{ij}^2$. Then, the marginal distribution of each W_i is χ_m^2 i.e., gamma with shape and scale parameters $m/2$ and $1/2$, and the correlation between W_i and $W_{i'}$ is $c_{i,i'}^2$. Finally, taking W_i/m to be the frailty for the ith patient yields a set of spatially correlated, individual-level, gamma-distributed frailties as required.

4.4   Point process models and geostatistics

Point process models are connected to geostatistics in two quite different ways. Firstly, the measurement process itself may be replaced by a point process. Sec-ondly, and as discussed in Chapter 1, in some applications the set of locations at which measurements are made should strictly be treated as a point process. This second aspect is usually ignored by making the analysis of the data con-ditional on the observed locations, although the conditioning is seldom made explicit. We now consider each of these two aspects in turn.

4.4.1 Cox processes

The essence of geostatistics, as distinct from other branches of spatial statistics, is that we wish to make inferences about a spatially continuous phenomenon, $S = \{S(x) : x \in \mathbb{R}^2\}$, which is not directly observable. Instead, we observe spatially discrete data, Y, which is stochastically related to S. By formulating a stochastic model for S and Y jointly and applying Bayes' Theorem we can, in principle, derive the conditional distribution of S given Y, and so use the observed data, Y, to make inferences about the unobserved phenomenon of scientific interest, S. Until now, in all of our models we have been able to represent Y as a vector $Y = (Y_1, \ldots, Y_n)$ in which each Y_i is associated with a location x_i, the Y_i are conditionally independent given S, and the conditional distribution of Y_i given S only depends on $S(x_i)$.

A Cox process (Cox, 1955) is a point process in which there is an unobserved, non-negative-valued stochastic process $S = \{S(x) : x \in \mathbb{R}^2\}$ such that, conditional on S, the observed point process is an inhomogeneous Poisson process with spatially varying intensity $S(x)$. Models of this kind fit into the general geostatistical framework whereby the model specifies the distributions of an unobserved spatial process S and of an observed set of data Y conditional on S, except that now the conditional distribution of Y given S is that of a Poisson process generating a random set of points $x_i \in \mathbb{R}^2$, rather than of a finite set of measurements Y_i at pre-specified locations x_i. The analogy is strengthened by the fact that the conditional Poisson process of Y given S is the point process analogue of mutually independent Y_i given S when each Y_i is a measured variable. Indeed, the Cox process can be derived as the limiting form of a geostatistical model of the following kind. Counts Y_i are observed at lattice points x_i with lattice-spacing δ. The Y_i are mutually independent Poisson-distributed random variables conditional on a real-valued, unobserved process S, with conditional expectations $\mu_i = \int S(x)dx$, where the integral is over the square of side δ centred on x_i. The limiting form of this model as $\delta \to 0$ is a Cox process.

One of the more tractable forms of Cox process is the log-Gaussian Cox process, in which $\log S$ is a Gaussian process (Møller et al., 1998). Brix and Diggle (2001) developed predictive inference for a spatio-temporal version of a log-Gaussian Cox process. Their motivation was to analyse data corresponding to the locations and times of individual cases of an acute disease, when the goal was to monitor temporal changes in the spatial variation of disease risk.

Diggle, Rowlingson and Su (2005) describe a specific application of this model in a spatio-temporal setting. They develop a real-time surveillance methodology in which the data consist of the locations and dates of all reported cases of non-specific gastroenteric illness in the county of Hampshire, UK. In this application, the spatio-temporal conditional intensity of the Cox process is modelled as

$$\lambda(x,t) = \lambda_0(x)\mu_0(t) \exp\{S(x,t)\}$$

where $\lambda_0(x)$ and $\mu_0(t)$ are deterministic functions which describe the long-term patterns of spatial and temporal variation in incidence, whilst $S(x,t)$ is a stationary Gaussian process which models spatially and temporally localised deviations from the long-term pattern. In a surveillance context, deviations of this kind potentially represent early warnings of "anomalies" in the data which may require further investigation. Hence, the statistical problem discussed in Diggle et al. (2005) is to predict $S(x, t)$ given the data on prevalent and incident cases, and in particular to identify places and times for which $\exp\{S(x, t)\}$ exceeds a pre-declared intervention threshold.

Brix and Møller (2001) and Benes, Bodlak, Møller and Waagepetersen (2001) also describe extensions of the log-Gaussian Cox process to spatio-temporal settings. In Brix and Møller (2001), the model is used to describe the invasion of a planted crop by weeds. In Benes et al. (2001), the application is to the mapping of spatial variations in disease risk when the locations of individual cases of the disease are known, a context very similar to the disease surveillance setting of Diggle et al. (2005).

Inference for the log-Gaussian Cox process generally requires computationally intensive Monte Carlo methods, whose implementation involves careful tuning. This applies in particular to likelihood-based parameter estimation (Møller and Waagepetersen, 2004) and to prediction of functionals of $S(\cdot)$ (Brix and Diggle, 2001). However, the following moment-based method provides an analogue of the variogram, which can be used for exploratory analysis and preliminary estimation of model parameters.

We assume that $S(\cdot)$ is stationary, and denote by μ and $\gamma(\cdot)$ its mean and covariance function. Then, the mean and covariance function of the intensity surface, $\Lambda(x) = \exp\{S(x)\}$, are $\lambda = \exp\{\mu + 0.5\gamma(0)\}$, which also represents the expected number of points per unit area in the Cox process, and $\phi(u) = \exp\{\gamma(u)\} - 1$. Now, define the reduced second moment measure of a stationary point process to be $K(s)$, where $\lambda K(s)$ is the expected number of further points within distance s of an arbitrary point of the process (Ripley, 1977). For the log-Gaussian Cox process the function $K(s)$ takes the form

$$K(s) = \pi s^2 + 2\pi\lambda^{-2} \int_0^s \phi(u)u\,du. \qquad (4.7)$$

A non-parametric estimator for $K(s)$, based on data consisting of n points x_i within a region A, is

$$\hat{K}(s) = \frac{|A|}{n(n-1)} \sum_{i=1}^n \sum_{j \neq i} w_{ij}^{-1} I(u_{ij} \leq s), \qquad (4.8)$$

where u_{ij} is the distance between x_i and x_j, $I(\cdot)$ is the indicator function, $|A|$ is the area of A and w_{ij} is the proportion of the circumference of the circle with centre x_i and radius u_{ij} which lies within A (Ripley, 1977). The estimator (4.8) essentially uses observed averages of counts within discs centred on each data-point x_i to estimate the corresponding theoretical expected count but with an edge-correction, represented by the w_{ij}, to adjust for the expected numbers of unobserved events at locations outside A. Preliminary estimates of model parameters can then be obtained by minimising a measure of the discrepancy between theoretical and empirical K-functions.

The K-function is widely used in the analysis of spatial point pattern data. For book-length discussions, see Diggle (2003) or Møller and Waagepetersen (2004).

4.4.2 Preferential sampling

A typical geostatistical data-set consists of a finite number of locations x_i and associated measurements Y_i. If, in this setting, we acknowledge that both the measurements and the locations are stochastic in nature, then a model for the data is a joint distribution for measurements and locations, which we represent formally as $[X, Y]$.

As discussed briefly in Section 1.2.3, we usually assume that sampling is non-preferential i.e., sampling and measurement processes are independent and the joint distribution of X and Y factorises as $[X, Y] = [X][Y]$. It follows that a conventional geostatistical analysis, by which we mean an analysis which conditions on X, is correctly targeted at the unconditional distribution of Y, and hence at the unconditional distribution of the underlying signal.

If, in contrast, sampling is preferential, then one of two possible factorisations of the joint distribution of X and Y is as $[X, Y] = [X][Y|X]$. Hence, the implicit inferential target of a conventional geostatistical analysis, which analyses only the data Y, is the conditional distribution $[Y|X]$, whereas the intended target is usually the unconditional distribution $[Y]$, and there is no reason in general to suppose that the two are equal.

It does not follow from the above argument that inferences which ignore preferential sampling will necessarily be badly misleading, but it does follow that we should be wary of accepting them uncritically. Provided that the model for $S(x)$ is known, standard kriging may still give reasonable results. Suppose, for example, that the stationary Gaussian model holds and that sampling favours locations x for which $S(x)$, and hence $Y_i = S(x_i) + Z_i$, is atypically large. The kriging predictor will then down-weight the individual influence of the large values of Y_i which would tend to occur in spatial concentrations within the over-sampled regions, and up-weight the influence of small, but spatially isolated, values of Y_i.

When, as is invariably the case in practice, model parameters are unknown, the consequences of ignoring preferential sampling are potentially more serious because standard methods of estimation will tend to produce biased estimates, which in turn will adversely affect the accuracy of predictive inferences concerning the signal. Again assuming that relatively large values are over-sampled, this would result in a positively biased estimate of the mean, and hence a tendency for predictions to be too large on average.

A model-based response to the preferential sampling problem is to formulate a suitable joint model for the response data Y and the locations X. The most natural way to do this is through their mutual dependence on the underlying signal process, $S = \{S(x) : x \in \mathbb{R}^2\}$. For example, we might first assume that, conditional on S, the measured values Y_i at locations x_i are mutually independent, $Y_i \sim N(S(x_i), \tau^2)$, as in the standard Gaussian linear model. A simple, if somewhat idealised, model for the preferential sampling mechanism might then be that, conditional on S, the sampled locations $X = (x_1, \ldots, x_n)$ are generated by a Poisson process with intensity $\lambda(x) = \exp\{\alpha + \beta S(x)\}$. Positive or negative β would correspond to over-sampling of large or small values of $S(x)$, respectively. To complete the model specification, the simplest assumption would be that S is a stationary Gaussian process. To emphasise that the locations at which we observe Y are determined by the point process X, we partition S as $S = \{S(X), S(\bar{X})\}$ where $\bar{X}$ denotes all locations which are not points of X. Then, the joint distribution of S, X and Y can be factorised as

$$[S, Y, X] = [S][X|S][Y|S(X)]. \tag{4.9}$$

In most geostatistical problems, the target for inference is $[S]$. The predictive distribution of S is $[S|Y, X] = [S, Y, X]/[Y, X]$, where $[Y, X]$ follows in principle from (4.9) by integration,

$$[Y, X] = \int [S, Y, X] dS,$$

although the integral may be difficult to evaluate in practice. Note that the conditional distribution $[Y|S(X)]$ in (4.9) is not of the standard form whereby the Y_i are mutually independent, $Y_i \sim N(S(x_i), \tau^2)$, because of the inter-dependence between S and X. We contrast (4.9) with the superficially similar model

$$[S, Y, X] = [S][X|S][Y|S] \tag{4.10}$$

where now $[Y|S]$ is a set of independent univariate Gaussian distributions. The model (4.10) would be appropriate if we observed a point process X and a set of measured values Y_i at pre-specified locations x_i, rather than at the points of X. This second situation is not without interest in its own right. It would arise, for example, if X represented a set of events whose spatial distribution is of scientific interest and were thought to depend on a spatially varying covariate $S(x)$ which is not directly observable everywhere but can be measured, possibly with error, at a set of pre-specified sample locations $x_i : i = 1, \ldots, n$. A specific example is considered by Rathbun (1996) in a study of the association between a point process of tree locations in a forest and an incomplete set of measured elevations.

Simulation results in Menezes (2005) confirm that when geostatistical data are generated from the model (4.9), standard geostatistical inferences which ignore the preferential sampling mechanism can be very misleading. Here, we give a single example to illustrate.

We simulated the signal process on a discrete grid of 100 by 100 points in a unit square, using a stationary Gaussian process with zero mean, unit variance, and Matérn correlation function with parameters $\kappa = 1.5$ and $\phi = 0.2$. Holding the signal process fixed, we then took three samples of values, denoted by Y_1, Y_2 and Y_3, which we refer to as *random*, *preferential* and *clustered*, respectively. Each sample consists of the values of the signal at a set of 100 sampling locations from the 100 by 100 grid, as follows. For Y_1, the sampling locations are an independent random sample of size 100 i.e., each of the 10,000 points in the

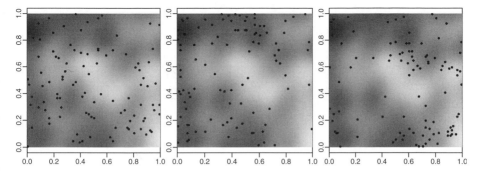

Figure 4.3. Sample locations and underlying realisations of the signal process for the example to illustrate the effects of preferential sampling. The left-hand panel shows the random sample, the centre panel the preferential sample and the right-hand panel the clustered sample. In each case, the grey-scale image represents the realisation of the signal process, $S(x)$, which was used to generate the associated measurement data.

Table 4.1. Sample statistics and parameter estimates for the three samples in the example to illustrate the effects of preferential sampling.

	Sampling statistics		Model parameter estimates		
Sample	Mean	Variance	$\hat{\mu}$	$\hat{\sigma}^2$	$\hat{\phi}$
Random	−0.13	0.42	0.2	0.86	0.21
Preferential	0.38	0.35	0.28	0.97	0.23
Clustered	−0.13	0.51	0.17	0.98	0.22

grid is equally likely to be selected. For Y_2, each grid-point x_i has probability of selection proportional to $\exp\{S(x_i)\}$ where $S(x_i)$ is the value of the signal at x_i. Finally, for Y_3 each point x_i has probability of selection proportional to $\exp\{S^*(x_i)\}$ where $S^*(x_i)$ is the simulated value at x_i of a second, independent realisation of the signal process. The samples Y_2 and Y_3 are spatially clustered to the same extent but Y_3, unlike Y_2, satisfies the standard geostatistical assumption that X and Y are independent.

Figure 4.3 shows the three samples of locations x_i together with the underlying realisation of the signal process. Note in particular that in the left-hand and right-hand panels, the pattern of the sample locations is unrelated to the spatial variation of the signal process.

For each of the three samples we obtained maximum likelihood estimates of the model parameters μ, σ^2 and ϕ, treating κ as known and, in the case of Y_2, ignoring the preferential nature of the sampling. Table 4.1 shows the maximum likelihood estimates together with the sample means and variances. The preferential sampling has a pronounced effect on the sample mean, as would be expected. In all three cases, the sample variance grossly under-estimates the variance of the signal process. The maximum likelihood estimates give reasonable results for all three model parameters except that, in the case of preferential sampling, there is still some indication of the biasing effect on the estimation of the mean.

Table 4.2. Mean square prediction errors for the three samples in the example to illustrate the effects of preferential sampling, using true and estimated parameter values.

	Random	Preferential	Clustered
True	0.0138	0.0325	0.0192
Estimated	0.0138	0.0326	0.0191

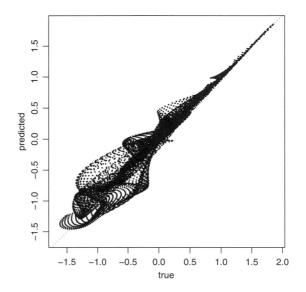

Figure 4.4. Predicted versus true values of the signal at 10,000 grid locations, using preferentially sampled data in conjunction with true values for all model parameters.

We then used each of the three samples to predict the signal at the original 10,000 grid locations, using both true and estimated parameter values. Table 4.2 gives the resulting average squared prediction errors. The larger values for the clustered than for the random sample illustrates that the former is a less efficient design for spatial prediction, whilst the preferential sample gives larger values still. Note also that using true parameters does not necessarily give a smaller averaged squared prediction error than using estimated values, because the estimated values reflect the characteristics of the particular realisation of the signal process. Finally, Figure 4.4 shows, for the preferential sample, a scatterplot of the 10,000 individual predictions against the true values of the signal, using true values of the model parameters for the predicted values. The preferential sample does a very good job of predicting the larger values of the signal, but is less reliable for smaller values, as a consequence of the under-sampling of sub-regions where the signal takes relatively small values.

Models for preferential sampling can also be considered as models for marked point processes. A marked point process is a point process, each of whose points has an associated random variable called the *mark* of the point in question. Marks may be qualitative or quantitative. In this context, it is not necessary for the mark to exist at every point in space, only at each point of the process, for example the points could be the locations of individual trees in a forest and the marks might denote the species (qualitative) or height (quantitative) of each tree. However, the marks could also be the values, at each point, of an underlying spatially continuous random field. In this case, the model in which the mark process is independent of the point process is called the random field model. The random field model for a marked point process is therefore the counterpart of non-preferential sampling for a geostatistical model. Schlather, Ribeiro Jr and Diggle (2004) consider methods for investigating the goodness-of-fit of the random field model to marked point process data.

4.5   Some examples of other model constructions

We have emphasised the role of the generalised linear model because it is widely useful in applications and is sufficiently general to introduce the main ideas of model-based geostatistics. However, it is clearly not universally applicable. To underline this, we give some simple examples of different model constructions, with suggestions for further reading.

4.5.1   Scan processes

A long-established method for sampling point process data *in situ* is quadrat sampling. This consists of counting the number of points of the process which fall within a demarcated spatial sampling unit, traditionally a square. A complete sample then consists of counts obtained from a series of quadrats placed randomly or systematically over the study region (Greig-Smith, 1952). In their original setting, quadrats would be placed at a discrete set of locations. Naus (1965) introduced the idea of scanning a point process with a continuously moving circular quadrat and using the maximum count as a way of testing for clustering in the underlying point process. Let $Y(x)$ denote the number of points of the process which lie within a fixed distance, r say, of x. Cressie (1993; chapter 5) called the process $N(x)$ a scan process.

Suppose that the underlying point process is a Cox process with intensity $S(x)$. Then, conditional on $S(\cdot)$, the observed count $Y(x)$ is Poisson-distributed with conditional expectation

$$\mu(x|S) = \int S(x - u)du,$$

where the integral is over a disc of radius u. This model is similar in some respects to a Poisson generalised linear geostatisical model, but is also different in at least one important respect, which is that observed counts cannot be conditionally independent given $S(\cdot)$; in fact, any realisation of $Y(x)$ will be piece-wise constant. A secondary consideration is that $S(x)$ cannot strictly be Gaussian, as it must be non-negative valued. Also, if we take $S(\cdot)$ to be log-Gaussian so as to meet the non-negative valued requirement, the log-Gaussian distribution is not preserved when we integrate $S(x)$ over a disc to obtain the conditional expectation of $Y(x)$.

4.5.2   Random sets

A random set (Matheron, 1971a) is a partition of a spatial region A into two sub-regions according to the presence or absence of a particular phenomenon, so defining a binary-valued stochastic process $S(x)$. A point process can be considered as a countable random set, but the term is usually applied to spatially continuous phenomena, for example a partition of a geographical area into land and water. A widely used model is the Boolean model (Serra, 1980), in which the random set is constructed as the union of a basic set, such as a disc, translated to each of the points of a homogenous Poisson process.

   Random sets have developed an extensive theory and methodology in their own right. Matheron (1971a) is an early account of a theory of random sets. Serra (1982) is a detailed account of theory and methods. A very extensive body of work under the heading of stereology is concerned essentially with the analysis of random sets in three spatial dimensions which are sampled using two-dimensonal sections or one-dimensional probes (Baddeley and Vedel Jensen, 2005). For further discussion and references, see also Cressie (1993, chapter 9) or Chilès and Delfiner (1999, section 7.8).

4.6   Computation

4.6.1   Simulating from the generalised linear model

Poisson model

Below, we give the sequence of commands for simulating from the Poisson log-linear model as shown in Figure 4.5. We first define the object cp to contain the coordinates of the required data locations. Next we use the function grf() to simulate a realisation of the Gaussian process at these locations with $\mu = 0.5$, $\sigma^2 = 2$ and Matérn correlation function with $\kappa = 1.5$, $\phi = 0.2$. We then store the Gaussian data in the object s; in Figure 4.5, these values are represented by the grey-scale shading of the grid squares. Next, we exponentiate the realised values of the Gaussian process to define the Poisson means. These are then passed to the function rpois() to simulate the conditionally independent Poisson counts. The simulated counts are indicated by the numbers shown in Figure 4.5. The spatially discrete representation of the underlying signal S in Figure 4.5 gives an alternative way of visualising the simulated data, instead of the superposition of a contour plot and a grey-scale image as used in Figure 4.1.

```
> set.seed(371)
> cp <- expand.grid(seq(0, 1, l = 10), seq(0, 1, l = 10))
> s <- grf(grid = cp, cov.pars = c(2, 0.2), cov.model = "mat",
+      kappa = 1.5)
> image(s, col = gray(seq(1, 0.2, l = 21)))
> lambda <- exp(0.5 + s$data)
> y <- rpois(length(s$data), lambda = lambda)
> text(cp[, 1], cp[, 2], y, cex = 1.5)
```

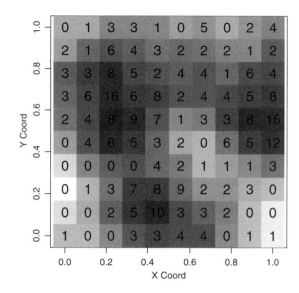

Figure 4.5. A simulation of the Poisson log-linear model. The numbers are the Poisson counts corresponding to locations at the centre of each grid square. The grey-scale represents the value of the underlying Gaussian process at each location.

The simulation model can be extended in various ways. For example, to include in the simulation non-spatial extra-Poisson variation of the kind discussed at the end of Section 4.3.1, we simply replace the command `lambda <- exp(s$data)` above by

```
> lambda <- exp(s$data + tau * rnorm(length(s$data)))
```

The additional term within the exponential generates independent Gaussian deviates with zero mean and variance τ^2, which are added to the values of the underlying Gaussian process. Similarly, to include a spatially varying mean, we would add a regression term within the exponential.

Bernoulli model

Below, we give the code for the simulation shown in Figure 4.2. For better visualisation the underlying Gaussian process is simulated at 401 locations equally spaced in the unit interval and the logit transformation is applied at each location to obtain the corresponding conditional probabilities. The object ind is then used to select 51 equally spaced points, and the binary values at these selected locations are generated using the rbinom() function.

```
> set.seed(34)
> locs <- seq(0, 1, l = 401)
> s <- grf(grid = cbind(locs, 1), cov.pars = c(5, 0.1),
+     cov.model = "matern", kappa = 1.5)
> p <- exp(s$data)/(1 + exp(s$data))
> ind <- seq(1, 401, by = 8)
```

```
> y <- rbinom(p[ind], size = 1, prob = p)
> plot(locs[ind], y, xlab = "locations", ylab = "data")
> lines(locs, p)
```

Binomial model

The 60 numbers shown in Figure 4.6 are simulated from a model with $[Y(x)|S] \sim Bin\{n, p(x)\}$ with $n = 5$ and $p(x) = \exp\{\mu + S(x)\}/[1 + \exp\{\mu + S(x)\}]$, where $S(x)$ is a Gaussian process with mean $\mu = 2$ and Matérn correlation function with $\kappa = 1.5$, $\phi = 0.15$. The circles in Figure 4.6 are drawn with radii proportional to the corresponding values of the underlying Gaussian process. To generate this simulation we first simulate from the Gaussian model, then logit-transform the simulated values to obtain the probabilities which we use to simulate the binomial data. A method for the function `points()` plots the Gaussian values. Finally, we use the standard R function `text()` to show the simulated binomial data as numbers above each sampling location. Our purpose in showing Figure 4.6 is not specifically to recommend this form of display, but more to illustrate different possibilities for visualisation of spatial data. The current example is one instance in which colour might be particularly effective, for example by using the radius of each circle to represent the corresponding realised value of the underlying Gaussian process and a discrete colour code for the actual count.

```
> set.seed(23)
> s <- grf(60, cov.pars = c(5, 0.25))
> p <- exp(2 + s$data)/(1 + exp(2 + s$data))
> y <- rbinom(length(p), size = 5, prob = p)
> points(s)
> text(s$coords, label = y, pos = 3, offset = 0.3)
```

In all of these examples, it is instructive to repeat the simulations with different values of the model parameters so as to gain insight into how details of the model specification do or do not affect the appearance of the simulated realisations. Replicate simulations holding parameter values constant similarly give useful insights into the behaviour of the models.

4.6.2 Preferential sampling

Next we show how to simulate random, preferential and clustered samples as used in the example of Section 4.4.2. First, we simulate the signal $S(x)$ in a grid of 10,000 points using `grf()`. Next we obtain measurements Y_i corresponding to 50 points sampled at random using `sample.geodata()`, which are returned as the `geodata` object `yr`. Note that there is no nugget term in this example, hence the sampled measurements are $Y_i = S(x_i)$, where x_i is the ith sampled location.

To simulate the preferential sample we make the probability that any point k from the grid is sampled proportional to $\exp\{bS_k\}$, where b in the example below is 1.2 and `S$data` is the simulated value of the signal at the kth grid-

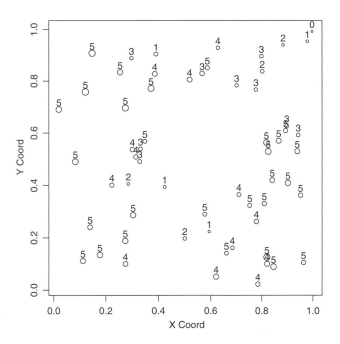

Figure 4.6. Simulated binomial data. Circles are drawn at the data locations, with radii proportional to the corresponding values of the underlying Gaussian process. Binomial counts are shown as numbers above the corresponding circles.

point. The sampled values $Y_i = S(x_i)$ are now returned as the geodata object yp.

Finally, to simulate a clustered sample we first generate a second, independent realisation of the signal process, $S_2(x)$ say, and make the probability of sampling point k from the grid proportional to $\exp\{bS_{2k}\}$ with sampled measurements in yc.

```
> set.seed(2391)
> S <- grf(10000, grid = "reg", cov.pars = c(1, 0.2))
> yr <- sample.geodata(S, size = 50)
> yp <- sample.geodata(S, size = 50, prob = exp(1.2 * S$data))
> S2 <- grf(10000, grid = "reg", cov.pars = c(1, 0.2))
> yc <- sample.geodata(S, size = 50, prob = exp(1.2 * S2$data))
```

4.7 Exercises

4.1. Investigate the consequence of using a quadratic, rather than a linear, Taylor series approximation to the function $g(\cdot)$ in the derivation of an approximate expression for the variogram of a GLGM.

4.2. Obtain an expression for the variogram of a Poisson log-linear model in which measurements $Y_i : i = 1, \ldots, n$ at locations x_i are conditionally independent, Poisson-distributed with conditional expectations μ_i, where $\log \mu_i = \alpha + S(x_i) + Z_i$, $S(\cdot)$ is a mean-square continuous stationary Gaussian process and $Z_i : i = 1, \ldots, n$ are mutually independent $N(0, \tau^2)$. Compare your general result with the special case $\tau^2 = 0$ and comment.

4.3. Consider the non-spatial GLMM in which counts $Y_i : i = 1, \ldots, n$ are conditionally independent, Poisson-distributed with conditional expectations $\mu_i = \exp(\alpha + Z_i)$, where $Z_i : i = 1, \ldots, n$ are mutually independent $N(0, \tau^2)$. Obtain the minimum mean square error predictors of the μ_i and their associated prediction variances. Investigate how these quantities depend on n, α and τ^2. Comment on the implications for spatial prediction using a Poisson log-linear GLGM.

4.4. Write code to simulate binomial geostatistical data with varying binomial denominators at the different sample locations. Experiment with alternative forms of visualisation for data of this kind.

5

Classical parameter estimation

In this chapter, we discuss methods for formulating a suitable geostatistical model and estimating its parameters. We use the description "classical" in two different senses: firstly, as a reference to the variogram-based methods of estimation which are widely used in classical geostatistics as developed by the Fontainebleau school; secondly, within mainstream statistical methodology as a synonym for non-Bayesian. The chapter has a strong focus on the linear Gaussian model. This is partly because the Gaussian model is, from our perspective, implicit in much of classical geostatistical methodology, and partly because model-based estimation methods are most easily implemented in the linear Gaussian case. We discuss non-Bayesian estimation for generalized linear geostatistical models in Section 5.5, indicating in particular why maximum likelihood estimation is feasible in principle, but difficult to implement in practice.

As discussed in Chapter 2, formulating a model for a particular application involves both spatial and non-spatial exploratory analysis. Our starting point for the remainder of this chapter is that we have identified a candidate model for data $Y_i : i = 1, \ldots, n$ observed at spatial locations $x_i : i = 1, \ldots, n$, with a mean structure $E[Y_i] = \mu_i$ and whose covariance structure is to be determined. Also, we assume that $\mu_i = \mu(x_i)$ where

$$\mu(x) = \beta_0 + \sum_{j=1}^{p} \beta_j d_j(x) \qquad (5.1)$$

and the $d_j(x)$ are spatial explanatory variables. From a model-based perspective, the mean and covariance structure together define a linear Gaussian model for the data; from a classical geostatistical perspective the mean and covariance structure define a model, but with no implication that the data follow a Gaussian distribution.

5.1   Trend estimation

For initial estimation of the mean parameters β_j, we use an ordinary least squares criterion, choosing estimates $\tilde{\beta}_j$ to minimise the quantity

$$RSS(\beta) = \sum_{i=1}^{n} (Y_i - \mu_i)^2. \tag{5.2}$$

At this point, it is helpful to use standard matrix notation for the linear model. Let $Y = (Y_1, \ldots, Y_n)$, $\beta = (\beta_1, \ldots, \beta_p)$ and write D for the $n \times (p + 1)$ matrix with an initial column of ones, and remaining columns containing the values of the explanatory variables $d_j(x_i) : i = 1, \ldots, n$. Then, the estimates $\tilde{\beta}$ which minimise (5.2) are

$$\tilde{\beta} = (D'D)^{-1}D'Y. \tag{5.3}$$

Assuming that the model for the mean has been correctly specified, the resulting estimates are unbiased, irrespective of the covariance structure. If we knew the covariance matrix of Y, say V, then a more efficient estimate would be the generalized least squares estimate,

$$\hat{\beta} = (D'V^{-1}D)^{-1}D'V^{-1}Y. \tag{5.4}$$

The estimate $\hat{\beta}$ is again unbiased, but also has the smallest variance amongst all unbiased linear estimates, $\beta^* = AY$. If we also assume that Y has a multivariate Gaussian distribution, then $\hat{\beta}$ is the maximum likelihood estimate.

Having obtained estimates $\tilde{\beta}$, we define the (ordinary least squares) residuals $R_i : i = 1, \ldots, n$ as the elements of the vector

$$R = Y - D\tilde{\beta}. \tag{5.5}$$

As discussed in Section 5.2 below, we use the residuals to identify a suitable parametric model for the covariance structure and to obtain initial estimates of covariance parameters. In later sections we then discuss how to refine our initial parameter estimates for both the mean and covariance structure and to make formal inferences.

5.2   Variograms

In Chapters 2 and 3, respectively, we introduced the *empirical* and *theoretical* variogram. We now re-visit and extend the earlier discussion.

5.2.1   The theoretical variogram

Recall from Section 3.1, equation (3.1) that the *theoretical variogram* of a spatial stochastic process is the function

$$V(x, x') = \frac{1}{2}\text{Var}\{S(x) - S(x')\}.$$

For a stationary or intrinsic process, the variogram reduces to a function of $u = ||x - x'||$. The second-moment properties of a stationary stochastic process $S(x)$ can therefore be described either by its *covariance function*, $\gamma(u) = \text{Cov}\{S(x), S(x - u)\}$, or its *variogram*, $V(u) = \frac{1}{2}\text{Var}\{S(x) - S(x - u)\}$. Their equivalence is expressed by the relation $V(u) = \gamma(0) - \gamma(u) = \sigma^2\{1 - \rho(u)\}$, where $\sigma^2 = \text{Var}\{S(x)\}$ and $\rho(u) = \text{Corr}\{S(x), S(x - u)\}$.

Because the mean of a stationary process is constant, the variogram in the stationary case can also be defined as $V(u) = \frac{1}{2}\text{E}[\{S(x) - S(x - u)\}^2]$. Now, suppose that the data $(x_i, y_i) : i = 1, \ldots, n$ are generated by a stationary process

$$Y_i = S(x_i) + Z_i$$

where Z_i are mutually independent, identically distributed with zero mean and variance τ^2. We define the variogram of the observation process, $V_Y(u)$ say, by

$$V_Y(u_{ij}) = \frac{1}{2}\text{E}[(Y_i - Y_j)^2]$$

where $u_{ij} = ||x_i - x_j||$. It follows that

$$V_Y(u) = \tau^2 + \sigma^2\{1 - \rho(u)\}. \tag{5.6}$$

Typically, $\rho(u)$ is a monotone decreasing function with $\rho(0) = 1$ and $\rho(u) \to 0$ as $u \to \infty$. In these circumstances, equation (5.6) neatly summarises the essential qualities of a classical geostatistical model. The typical variogram is a monotone increasing function with the following features. The intercept, τ^2, corresponds to the *nugget* variance. The asymptote, $\tau^2 + \sigma^2$, corresponds to the variance of the observation process Y, sometimes called the *sill*, which in turn is the sum of the nugget variance and the signal variance, σ^2. The way in which the variogram increases from its intercept to its asymptote is determined by the correlation function $\rho(u)$, the most important features of which are its behaviour near $u = 0$, which relates to the analytic smoothness of the underlying signal process, and how quickly $\rho(u)$ approaches zero with increasing u, which reflects the physical extent of the spatial correlation in the process. When $\rho(u) = 0$ for u greater than some finite value, this value is known as the *range* of the variogram. When $\rho(u)$ only approaches zero asymptotically as u increases, the range is undefined. We then follow geostatistical convention by defining the *practical range* as the distance u_0 at which $\rho(u_0) = 0.05$, hence $V_Y(u_0) = \tau^2 + 0.95\sigma^2$. See Figure 5.1 for a schematic illustration.

The nugget variance, which in the current context equates to the intercept of $V_Y(u)$, is an important parameter for spatial prediction. As we will see in Chapter 6, the value of τ^2 affects the degree to which the predicted surface $\hat{S}(x)$ will track the observed data Y_i. In particular, setting $\tau^2 = 0$ will force spatial predictions to interpolate the data. A decision on whether to set $\tau^2 = 0$, or to estimate a positive value of τ^2, is therefore an important one when choosing the model family.

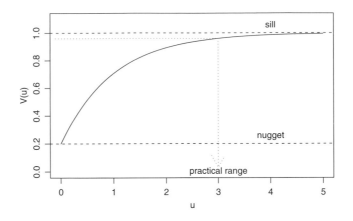

Figure 5.1. Schematic representation of a typical variogram, with structural parameters indicated.

5.2.2   The empirical variogram

From a data-analytic perspective, the second definition of the theoretical variogram as an expectation, $V_Y(u) = \frac{1}{2}\mathrm{E}[(Y_i - Y_j)^2]$, is important because it implies that, under the stationarity assumption, the observed quantities $v_{ij} = \frac{1}{2}(Y_i - Y_j)^2$ are unbiased estimates of the corresponding variogram ordinates, $V_Y(u_{ij})$. Note that some authors describe the variogram ordinates as "semivariances." The collection of pairs of distances and their corresponding variogram ordinates $(u_{ij}, v_{ij}) : j > i$ is called the *empirical variogram* of the data $(x_i, Y_i) : i = 1, \ldots, n$. The left-hand panel of Figure 5.2 shows the empirical variogram of the surface elevation data from Example 1.1 as a scatterplot. A plot of this kind is also called a *variogram cloud*. The extensive scatter in Figure 5.2 is typical, and severely limits the value of the empirical variogram as a data-analytic tool. The theoretical explanation for this is twofold. Firstly, under Gaussian modelling assumptions the marginal sampling distribution of each variogram ordinate v_{ij} is proportional to chi-squared on 1 degree of freedom, a highly skewed distribution with coefficient of variation $\sqrt{2} \approx 1.4$. Secondly, the empirical variogram ordinates are necessarily correlated because the $\frac{1}{2}n(n-1)$ distinct ordinates v_{ij} are derived from only n observations Y_i.

5.2.3   Smoothing the empirical variogram

To improve the behaviour of the empirical variogram as an estimator for the underlying theoretical variogram $V_Y(u)$, we need to apply some kind of smoothing. The rationale for so doing is that $V_Y(u)$ is expected to be a smoothly varying function of u, hence averaging values of v_{ij} over suitably narrow ranges of inter-point distances u_{ij} will reduce the variance without introducing material amounts of bias.

When the sample design is a regular lattice, the smoothing can be achieved without introducing any bias, simply by averaging all v_{ij} corresponding to each

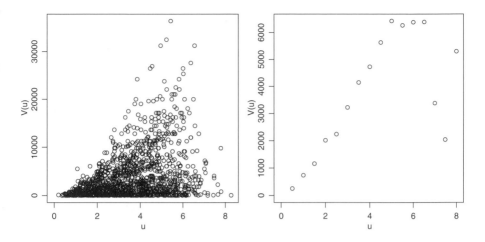

Figure 5.2. Variograms for the surface elevation data from Example 1.1: on the left the empirical variogram, and on the right the sample variogram using bin width $h = 0.5$.

distinct u_{ij}. For an irregular design, we compromise between variance and bias by averaging within declared ranges of u_{ij}. Hence, for bin width h we define *sample variogram* ordinates V_k, for positive integer k, as the averages of all v_{ij} for which the corresponding u_{ij} satisfy $(k-1)h < u_{ij} \le kh$. Then, V_k is an approximately unbiased estimate of $V_Y(u_k)$, where we adopt the convention that $u_k = (k - 0.5)h$, the mid-point of the corresponding interval. The exclusion of zero from the smallest of the binned intervals is deliberate. If the sample design includes duplicate measurements from coincident locations, the average of the corresponding empirical variogram ordinates v_{ij} provides a direct estimate of the nugget variance, τ^2, which can then be distinguished from small-scale spatial variation. In such cases, this estimate should be plotted as an additional point on the sample variogram. Otherwise, the nugget variance can only be estimated from the sample variogram by extrapolation.

The right-hand panel of Figure 5.2 shows the sample variogram of the surface elevation data from Figure 1.1, using a bin width of $h = 0.5$, or 25 feet. The first plotted ordinate, $\hat{V}(0.5)$, is close to zero, suggesting that the nugget variance is small i.e., elevation is measured with negligible error. The rising curve of sample variogram ordinates, levelling out at a distance of around $u \approx 5$ (250 feet), corresponds to a positive spatial correlation decaying with distance. The wild fluctuations in the sample variogram ordinates at large distances u are not untypical. They arise primarily because the empirical variogram ordinates are correlated, and the effects of this are more pronounced at relatively large distances. Also, as can be seen from the left-hand panel of Figure 5.2, the numbers of pairs of sample locations which contribute to the sample variogram ordinates diminish at very large distances. We therefore do not attach any particular significance to the large drop in the sample variogram ordinates beyond $u \approx 6$. For this reason, it is sensible to limit the sample variogram calculations to distances which are smaller than the maximum distance observed in the data. However, we are unable to offer an objective rule for what range of distances should be included; anticipating the discussion below, this is one reason why we are wary of using the sample variogram for formal inference.

More elaborate forms of smoothing of the empirical variogram are possible, for example using kernel or spline smoothers. The superficial justification for this is that estimating $V_Y(u)$ from the empirical variogram (u_{ij}, v_{ij}) is a nonparametric regression problem. The literature on smoothing methods for nonparametric regression is extensive. Accessible introductions include Bowman and Azzalini (1997). Recall, however, that the $\frac{1}{2}n(n-1)$ empirical variogram ordinates are not independent, nor do they have a common variance. For these reasons, conventional guidelines for nonparametric regression methods are inappropriate. Our view is that the sample variogram should be regarded primarily as a helpful initial display to identify broad features of the underlying covariance structure of the data, and also as a convenient way of obtaining initial estimates of model parameters, but not as a formal method of parametric inference. Hence, we would argue that subjective choice of band-width, in conjunction with a simple smoothing method such as binning by distance intervals, is sufficient. From this point of view, an important feature of the sample variogram of the elevation data is that its practical range is of the same order of magnitude as the dimension (6.7 units or 330 feet) of the study region. This led us in Chapter 2 to consider a non-stationary model for the data, incorporating a spatially varying mean.

5.2.4   Exploring directional effects

If directional effects are suspected, the scalar inter-point distances u_{ij} in the empirical variogram can be replaced by vector differences $x_i - x_j$ and the result displayed as a three-dimensional scatterplot; most modern software environments, including R, have facilities for dynamic graphical display of three-dimensional scatterplots. For the same reasons that the isotropic version of the empirical variogram is an ineffective data-analytic tool, this three-dimensional display is unlikely to reveal other than gross directional effects. However, we can apply the same binning method as in the isotropic case, for example, partitioning the space of vector differences into grid-cells and displaying the corresponding average variogram ordinates as a grey-scale image or contour plot. In practice, to achieve a useful level of detail in three-dimensional plots of this kind requires more data than for their two-dimensional counterparts. In this connection we again emphasise that we regard the sample variogram only as a helpful way of displaying the data prior to formal inference.

We again use the elevation data to illustrate this method of estimation, although these data are rather too sparse for a detailed exploration of directional effects. Nevertheless, the directional sample variogram, shown in Figure 5.3 as a contour plot, confirms that the spatial variation is substantially greater along the north-south axis than along the east-west axis. A glance at Figure 1.1 should convince the reader that this is a reasonable conclusion for these data. However, attributing this effect to a directional covariance structure is only one of several

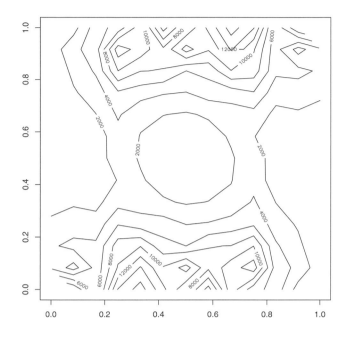

Figure 5.3. The directional sample variogram of the surface elevation data. The contour plot was constructed from the sample means of the empirical variogram ordinates in square bins of unit side.

possible explanations. Another is that the underlying process has a spatially varying mean, $\mu(x)$, as discussed in Chapter 2.

5.2.5 The interplay between trend and covariance structure

When the underlying mean function, $\mu(x)$, is not constant, empirical or sample variograms based on the observations Y_i are potentially very misleading. In this situation, the empirical variogram wrongly attributes the variation induced by the non-constant mean, $\mu(x)$, to large-scale covariance structure in the unobserved process $S(x)$. A solution is to estimate $\mu(x)$, typically by assuming either a trend surface model or, if covariate information is available, a more general regression model, and to convert the observations to residuals, $R_i = Y_i - \hat{\mu}(x_i)$, before calculating the empirical variogram. Of course, the properties of the observed residuals R_i do not exactly match those of the theoretical but unobserved residuals, $Y_i - \mu(x_i)$. However, their covariance structure should not differ too much from that of the true residuals provided the number of parameters estimated in $\hat{\mu}(x)$ is small relative to n, the number of observations.

As an illustration, we consider a simulation in which the sample design mimics that of the surface elevation data, but the simulated observations Y_i are

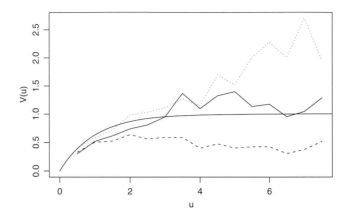

Figure 5.4. The sample variogram for a simulated data-set with a non-constant mean function. The theoretical variogram is shown as a smooth curve. The dotted line is the sample variogram based on uncorrected observations Y_i. The solid line is the sample variogram based on the true residuals, $Y_i - \mu_i$. The dashed line is the sample variogram based on the estimated residuals $Y_i - \hat{\mu}_i$.

generated by a model $Y_i = \mu(x_i) + S(x_i)$ in which $\mu(x)$ is a quadratic surface and $S(x)$ is a stationary Gaussian process with mean zero, variance $\sigma^2 = 1$ and exponential correlation function, $\rho(u) = \exp(-u)$. Figure 5.4 compares the theoretical variogram with sample variograms based on the observed values Y_i, the observed residuals $R_i = Y_i - \hat{\mu}(x_i)$ with mean parameters estimated by ordinary least squares, and the true residuals $R_i^* = Y_i - \mu_i$. The positive bias in the variogram based on the raw data Y_i arises from the non-stationary variation induced by the quadratic trend surface. Using either observed or true residuals produces estimates which are closer to the theoretical variogram. Note, however, that the sample variogram based on observed residuals lies below that based on true residuals. Because the observed residuals are defined so as to minimise the variation about the estimated mean, we might generally expect the sample variogram of observed residuals to exhibit negative bias. The discrepancy between observed and true residuals would be less marked in a larger data-set, and the negative bias in the sample variogram consequently smaller. This example illustrates how a decision on the data analyst's part to ascribe part of the spatial variation in a real data-set to a deterministic trend model can materially affect the results obtained in any subsequent estimation of spatial correlation structure.

When analysing real data, we have to make a subjective judgment as to whether we should remove an empirically estimated trend before estimating spatial correlation structure. Figure 5.5 illustrates the point using the surface elevation data. It shows the sample variogram of observed residuals after fitting a quadratic trend surface to the observed elevation values using ordinary least squares. If we compare this with the sample variogram of the unadjusted data, shown as the right-hand panel of Figure 5.2, we see a number of qualitative similarities: an intercept close to zero, a smooth rising trend approaching a plateau

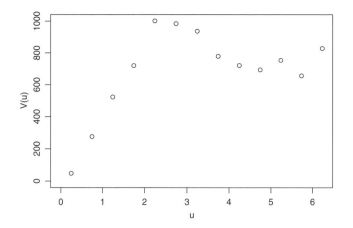

Figure 5.5. The sample variogram of observed residuals, after fitting a quadratic trend surface to the elevation data.

and erratic behaviour at large distances. However, the major differences are that the plateau is reached at smaller distances than before, $u \approx 2$ rather than $u \approx 5$, and its height is approximately 1000, whereas in Figure 5.2 the height of the plateau was approximately 6000. This shows that the fitted quadratic trend surface has accounted for approximately five-sixths of the total variation in the unadjusted elevation values, resulting in a weaker estimated spatial correlation structure for the residual variation than for the unadjusted elevations. For the time being, we regard these as alternative empirical descriptions of the pattern of spatial variation in the elevation data and make no attempt to say which, if either, is the better model in any scientific sense.

5.3   Curve-fitting methods for estimating covariance structure

In classical geostatistics, the variogram is used not only for exploratory purposes, but also for formal parameter estimation. In general we do not favour this approach, for reasons which we now discuss.

A possible rationale for using the variogram as the basis for parameter estimation is that the empirical variogram ordinates, v_{ij}, are unbiased estimates of the corresponding theoretical variogram ordinates, $V(u_{ij}; \theta)$, hence estimation of θ can be considered as an exercise in curve-fitting. In early work, the curve-fitting was often done "by eye," in other words by trying different values for the model parameters and visually inspecting the fit to the sample variogram. Although we do not advocate this as a method for parameter estimation, it can be a good way to find reasonable initial values for estimation methods involving numerical optimisation, which we discuss in the following sections. As discussed in Section 5.2.2, visual inspection of the empirical variogram is rarely helpful, and for curve-fitting by eye it is preferable to use the sample variogram.

More objective curve-fitting methods include the use of non-linear regression analysis, treating the empirical or sample variogram ordinate as the response variable and inter-point distance as the corresponding explanatory variable. In the following discussion of this more objective approach, we use the notation $(u_k, v_k, n_k) : k = 1, \ldots, m$ to denote a sample variogram. In this notation, v_k represents the averaged empirical variogram ordinates over the distance-bin with mid-point u_k, and n_k denotes the number of empirical variogram ordinates which contribute to v_k. The unsmoothed empirical variogram is the special case in which all $n_k = 1$. Rather than using the mid-point of the distance bin, a variation is to define u_k as the average of the inter-point distances which fall within the kth bin. As pointed out by a reviewer, this may be particularly appropriate when the empirical distribution of the inter-point distances is strongly multi-modal.

5.3.1   Ordinary least squares

The best-known objective curve-fitting algorithm is ordinary least squares. This estimates θ to minimise the criterion

$$S_0(\theta) = \sum_{k=1}^{m} \{v_k - V(u_k; \theta)\}^2. \tag{5.7}$$

An improvement, which recognises the effect of the varying n_k, is n-weighted least squares. The estimation criterion is now

$$S_n(\theta) = \sum_{k=1}^{m} n_k \{v_k - V(u_k; \theta)\}^2. \tag{5.8}$$

Note in particular that n-weighted least squares is almost equivalent to ordinary least squares applied to the empirical variogram; the two would be exactly equivalent if all of the u_k were exact distances between sampling locations as can be achieved, for example, with a lattice design. In practice, the efficiency of either (5.7) or (5.8) as a method of estimation depends on the choices of m and of the u_k.

5.3.2   Weighted least squares

Further refinements of the least squares method have been proposed, in response to the fact that the sampling variance of v_k depends on the corresponding value of the theoretical variogram, $V(u_k; \theta)$, as well as on n_k. Under Gaussian modelling assumptions, each empirical variogram ordinate v_{ij} has expectation $V(u_{ij}; \theta)$ and variance $2V(u_{ij}; \theta)^2$. This observation led Cressie (1985) to propose a V-weighted least squares estimation criterion,

$$S_V(\theta) = \sum_{k=1}^{m} n_k [\{v_k - V(u_k; \theta)\}/V(u_k; \theta)]^2. \tag{5.9}$$

As shown in unpublished work by Barry, Crowder and Diggle (1997), this corresponds to the use of a biased estimating equation, essentially because the unknown parameter, θ, contributes to the weighting. To see this, we differentiate $S_V(\theta)$ with respect to each element of θ. This gives, for each j,

$$\frac{\partial}{\partial \theta_j} S_V(\theta) = \sum_{k=1}^{m} 2n_k \left\{ \frac{v_k - V(u_k; \theta)}{V(u_k; \theta)} \times \left(\frac{-v_k}{V(u_k; \theta)^2} \right) \times \frac{\partial}{\partial \theta_j} V(u_k; \theta) \right\}$$

$$= \sum_{k=1}^{m} 2n_k \left\{ \frac{-v_k^2 + v_k V(u_k; \theta)}{V(u_k; \theta)^3} \frac{\partial}{\partial \theta_j} V(u_k; \theta) \right\}.$$

The V-weighted least squares estimates satisfy the estimating equations $D_j(\theta) = 0$ for all j, where

$$D_j(\theta) = \frac{\partial}{\partial \theta_j} S_V(\theta).$$

Since v_k is approximately unbiased for $V(u_k; \theta)$, it follows that

$$E[D_j(\theta)] \approx \sum_{k=1}^{m} 2n_k \left[-\frac{\mathrm{Var}(v_k)}{V(u_k; \theta)^3} \frac{\partial}{\partial \theta_j} V(u_k; \theta) \right] \neq 0, \qquad (5.10)$$

hence the estimating equations are biased. An intuitive explanation is that minimisation of (5.9) is equivalent to maximisation of a Gaussian likelihood but ignoring the determinant of the variance matrix. However, $\mathrm{Var}(v_k)$ is of order n_k^{-1} and for a given sample size n, the number of bins m is of order $\bar{n}_k^{-1}$ where $\bar{n}_k$ is the average of the n_k. Hence, (5.10) also suggests that in practice the amount of bias will decrease as the n_k increase. This result provides a theoretical justification for the practical guidelines given in applied geostatistics texts. See, for example, the recommendations in Journel and Huijbreghts (1978, pp. 193–194). Theoretical calculations of the effects of increasing the n_k can be made under either of two different conditions. The first, called *in-fill asymptotics*, envisages an increasing number of sample locations within a fixed spatial region. In contrast, *increasing domain asymptotics* envisages a constant density of sample locations in a region of increasing size. Note that under in-fill asymptotics, we can never achieve consistent parameter estimation because in general, observing a noise-free process $S(x)$ throughout a continuous spatial region does not determine its parameter values exactly. Under increasing domain asymptotics, we can hold a chosen set of distance-bins fixed, leading to increases in all of the n_k as the study region grows in size and consistency becomes achievable. Nevertheless, for a given data-set, we can only increase the n_k by increasing the bin width, and this introduces a second kind of bias, which we call *smoothing bias*, because the theoretical variogram varies non-linearly over the ranges of distances included within individual bins. Muller (1999) considered the special case of the empirical variogram, for which the bias in the implicit estimating equations is most pronounced, and showed that in this case the estimand when using (5.9) is $3V(u; \theta)$ rather than $V(u; \theta)$ itself.

An unbiased set of estimating equations could be obtained from an iteratively weighted least squares algorithm, as used in generalized linear modelling (McCullagh and Nelder, 1989). The resulting set of estimating equations, also given in Barry et al. (1997), solves $D_j^*(\theta) = 0$ for all j, where now

$$D_j^*(\theta) = \sum_{i=1}^{m} n_k \left\{ \frac{v_k - V(u_k; \theta)}{V(u_k; \theta)} \frac{\partial}{\partial \theta_j} V(u_k; \theta) \right\} \qquad (5.11)$$

and $E[D_j^*(\theta)] \approx 0$, as required. Similar ideas have been suggested by several authors, including Cressie (1985), McBratney and Webster (1986), Fedorov (1989) and Zimmerman and Zimmerman (1991), but appear not to have been widely used in practice.

Our first conclusion from the above discussion is that n-weighted least squares applied to the sample variogram gives a simple and convenient method for obtaining initial estimates of variogram parameters. Our second conclusion is that the more elaborately weighted criterion (5.9) is theoretically flawed, and we cannot therefore recommend it. The use of iteratively weighted least squares overcomes the specific theoretical objection to (5.9) by using an unbiased estimating equation but, as discussed in Section 5.3.3 below, our wider conclusion is that the variogram should be used as a graphical method of exploratory data analysis, rather than as a vehicle for formal parameter estimation.

5.3.3   Comments on curve-fitting methods

Other curve-fitting methods have been proposed. For example, Cressie and Hawkins (1980) suggest a criterion based on absolute differences between v_k and $V(u_k; \theta)$, which is less susceptible than are least squares criteria to outlying observations. Note in this context that for a sample design of n points, a single outlier amongst the y_i potentially contaminates $n-1$ empirical variogram ordinates.

In our opinion, the analogy between curve-fitting methods for variogram parameter estimation and non-linear regression modelling is a poor one, because of the inherent correlations amongst empirical variogram ordinates. One consequence of this correlation is that sample variograms often appear smooth, suggesting more precise estimation of the underlying theoretical variogram than is in fact the case. As an illustration of this effect, Figure 5.6 shows three sample variograms, generated from independent realisations of a Gaussian process with theoretical variogram $V(u) = 1 - \exp(-u/0.25)$ and $n = 100$ sample locations randomly distributed over a unit square region. Each of the three sample variograms presents a smooth curve, but their inherent imprecision as estimates of $V(u)$ is clearly shown by the wide divergence amongst the three realisations. The introduction of a more-or-less arbitrary bin width parameter, and of an upper limit for the range of distances to be included, are also unattractive features of what should be an objective procedure when formal parameter estimation, as opposed to informal exploratory analysis, is the goal.

Unless the sample design includes duplicate measurements at the same location, estimation of the intercept from the sample variogram involves extrapolation, which is always a dangerous exercise. To emphasise this, Figure 5.7 shows a sample variogram with two fitted theoretical variograms, each of which fits the sample variogram equally well, but with very different extrapolations

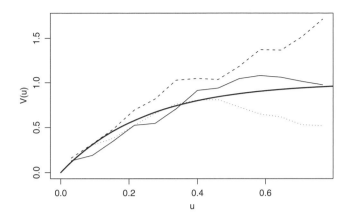

Figure 5.6. Sample variograms estimated from three independent realisations of the same stationary Gaussian process. The theoretical variogram model is shown as a smooth bold line.

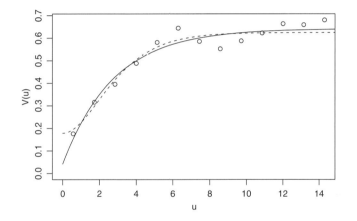

Figure 5.7. A sample variogram (small circles) and two theoretical variograms (solid and dashed lines) which appear to fit the data equally well.

to zero. This gives one reason for including τ^2 in the model routinely. Another is that in practice the nugget arises through a combination of measurement error and spatial variation on scales smaller than the smallest distance between non-coincident locations in the sample design. When fitting a parametric model we may therefore choose to compromise between the fit at $u = 0$ and the fit at small, positive values of u. But duplicate design points are still extremely helpful, and are necessary if we do want to preserve a formal distinction between pure measurement error and small-scale spatial variation.

Within the model-based paradigm, variogram-based parameter estimation is also inherently inefficient. Rather than estimate parameters directly from the variogram, which is only one of a number of possible summaries of the data, we prefer to declare an explicit model for the original data, $(x_i, y_i) : i = 1, \ldots, n,$

and to apply generally accepted principles of statistical estimation. This leads us to favour estimation methods based on the likelihood function. A legitimate concern with using likelihood-based methods is that they require additional distributional assumptions to be made about the data generating process. In particular, for continuous measurement data our approach will require us to assume that a Gaussian model is appropriate, either for the original data or after transformation. This places an increased emphasis on the need for diagnostic checking.

5.4   Maximum likelihood estimation

5.4.1   General ideas

Maximum likelihood estimation is a widely accepted statistical method, with well-known optimality properties in large samples. Under mild regularity conditions (Cox and Hinkley, 1974), the maximum likelihood estimator is asymptotically normally distributed, unbiased and fully efficient. Within the geostatistical context, implementation of maximum likelihood estimation is only straightforward when the data are generated by a Gaussian model. However, and with the added flexibility provided by marginal transformations of the response variable Y, this model is useful for many geostatistical applications in which Y is a continuous-valued quantity. Furthermore, we emphasise that the obstacles to implementation of maximum likelihood estimation in non-Gaussian models are only computational ones. The large-sample optimality properties of maximum likelihood estimation hold much more generally than in the Gaussian setting.

For general discussions of likelihood-based methods of statistical inference, including derivations of the results quoted above, we refer the reader to Cox and Hinkley (1974), Azzalini (1996) or Pawitan (2001).

5.4.2   Gaussian models

We shall consider the Gaussian model with a linear specification for the spatial trend, $\mu(x)$. This allows for the inclusion of a polynomial trend surface or, more generally, spatially referenced covariates. Hence for $\mu(x) = D\beta$,

$$Y \sim \mathrm{N}(D\beta, \sigma^2 R(\phi) + \tau^2 I) \tag{5.12}$$

where D is an $n \times p$ matrix of covariates, β is the corresponding vector of regression parameters, and R depends on a scalar or vector-valued parameter ϕ. The log-likelihood function is

$$\begin{aligned}
L(\beta, \tau^2, \sigma^2, \phi) = & -0.5\{n\log(2\pi) + \log\{|(\sigma^2 R(\phi) + \tau^2 I)|\} \\
& + (y - D\beta)^{\mathrm{T}}(\sigma^2 R(\phi) + \tau^2 I)^{-1}(y - D\beta)\},
\end{aligned} \tag{5.13}$$

maximisation of which yields the maximum likelihood estimates of the model parameters.

An algorithm for maximisation of the log-likelihood proceeds as follows. Firstly, we parameterise to $\nu^2 = \tau^2/\sigma^2$ and write $V = R(\phi) + \nu^2 I$. Given V, the log-likelihood function is maximised at

$$\hat{\beta}(V) = (D^{\mathrm{T}} V^{-1} D)^{-1} D^{\mathrm{T}} V^{-1} y \qquad (5.14)$$

and

$$\hat{\sigma}^2(V) = n^{-1} \{y - D\hat{\beta}(V)\}^{\mathrm{T}} V^{-1} \{y - D\hat{\beta}(V)\}. \qquad (5.15)$$

Note that $\hat{\beta}(V)$ reduces to the generalized least squares estimate (5.4) if V is known, rather than being a function of unknown parameters.

By substituting the above expressions for $\hat{\beta}(V)$ and $\hat{\sigma}^2(V)$ into the log-likelihood function, we obtain a concentrated log-likelihood

$$L_0(\nu^2, \phi) = -0.5\{n \log(2\pi) + n \log \hat{\sigma}^2(V) + \log |V| + n\}. \qquad (5.16)$$

This must then be optimised numerically with respect to ϕ and ν, followed by back substitution to obtain $\hat{\sigma}^2$ and $\hat{\beta}$.

The practical details of the optimisation may depend on the particular family under consideration. For example, when using the Matérn correlation function, our experience has been that the shape parameter κ is often poorly identified. Our preference is therefore to choose the value of κ from a discrete set, for example $\{0.5, 1.5, 2.5\}$, to cover different degrees of mean-square differentiability of the underlying signal process, rather than attempting to optimise over all positive values of κ.

Note also that different parameterisations of V may affect the convergence of the numerical optimisation. In particular, as discussed in Chapter 3 our standard parameterisation of the Matérn and powered exponential families leads to a natural interpretation as a scale parameter ϕ and a shape parameter κ, but the two parameters are not orthogonal in their effects on the induced covariance structure. As a consequence, neither are they orthogonal in the statistical sense; the maximum likelihood estimators for ϕ and κ tend to be strongly correlated. As discussed above, one response to this is to consider only a small number of candidate values for κ, corresponding to qualitatively different smoothness properties of the signal process. Another is to use the re-parameterisation suggested by Handcock and Wallis (1994), in which ϕ is replaced by $\alpha = 2\kappa^{0.5}\phi$. Zhang (2004) investigates the re-parameterisation question in detail, and shows that difficulties can also arise with respect to the signal variance, σ^2. Zhang's results demonstrate that, in a Matérn model with parameters σ^2, ϕ and known $\kappa = 0.5$, the ratio σ^2/ϕ is much more stably estimated than either σ^2 or ϕ themselves.

Re-parameterisation affects not only the performance of numerical optimisation algorithms, but also the adequacy of standard asymptotic approximations to the sampling distributions of maximum likelihood estimates. In personal communication, Zhang has suggested that better agreement between the finite-sample properties of maximum likelihood estimators and their asymptotic approximations is obtained by using a re-parameterisation to $\theta_1 = \log(\sigma^2/\phi^{2\kappa})$ and $\theta_2 = \log(\phi^{2\kappa})$, again treating κ as known. Our general experience has been that quadratic approximations to the log-likelihood surface are often poor, and standard errors derived by inverting a numerically estimated Hessian matrix can be unreliable. Also, we have found that, for example, estimating all three parameters in the Matérn model is very difficult because the parameters are poorly identified, leading to ridges or plateaus in the log-likelihood surface. Warnes and Ripley (1987) show an example of this phenomenon; see also the discussion in Stein (1999, pp. 172–173). For these reasons, we prefer to examine the behaviour of the log-likelihood surface by profiling, as we now describe.

5.4.3 Profile likelihood

In principle, the variability of maximum likelihood estimates can be investigated by inspection of the log-likelihood surface. However, the typical dimension of this surface does not allow direct inspection.

Another generic likelihood-based idea which is useful in this situation is that of profile likelihood. Suppose, in general, that we have a model with parameters (α, ψ) and denote its likelihood by $L(\alpha, \psi)$. We define the *profile log-likelihood* for α by

$$L_p(\alpha) = L(\alpha, \hat{\psi}(\alpha)) = \max_{\psi}(L(\alpha, \psi)).$$

In other words, we consider how the likelihood varies with respect to α when, for each value of α, we assign to ψ the value which maximises the log-likelihood with α held fixed. The profile log-likelihood allows us to inspect a likelihood surface for α, which is of lower dimension than the full likelihood surface. It can also be used to calculate approximate confidence intervals for individual parameters, exactly as in the case of the ordinary log-likelihood for a single parameter model (Cox and Hinkley, 1974). Note that the concentrated log-likelihood (5.16), which we introduced as a computational device for maximum likelihood estimation, can now be seen to be the profile log-likelihood surface for (ν^2, ϕ) in the model (5.12).

5.4.4 Application to the surface elevation data

We now apply the method of maximum likelihood to the surface elevation data. We adopt the Matérn family of correlation functions, and consider candidate values $\kappa = 0.5$, 1.5 and 2.5 for the shape parameter. We place no constraint on τ^2, although the context in which these data arise suggests that τ^2 should be relatively small.

We first fit the model under the assumption that the mean is constant. The left-hand panel of Figure 5.8 shows the resulting fitted variograms for each of $\kappa = 0.5$, 1.5 and 2.5, whilst the upper half of Table 5.1 gives the corresponding parameter estimates and maximised log-likelihoods. Visual inspection of Figure 5.8 and comparison of the maximised log-likelihoods suggest that $\kappa = 0.5$ gives a poor fit, whereas the fits for $\kappa = 1.5$ and 2.5 are comparable. Note in particular that the likelihood criterion leads to an estimated theoretical variogram which gives a good visual fit to the sample variogram at small distances u, but a less good fit at large distances. This illustrates how the likelihood cri-

Table 5.1. Parameter estimates for the surface elevation data for models with constant mean and a linear trend on the coordinates.

Model	$\hat{\mu}$			$\hat{\sigma}^2$	$\hat{\phi}$	$\hat{\tau}^2$	logL
			Model with constant mean				
$\kappa = 0.5$	863.71			4087.6	6.12	0	−244.6
$\kappa = 1.5$	848.32			3510.1	1.2	48.16	−242.1
$\kappa = 2.5$	844.63			3206.9	0.74	70.82	−242.33

Model	$\hat{\beta}_0$	$\hat{\beta}_1$	$\hat{\beta}_2$	$\hat{\sigma}^2$	$\hat{\phi}$	$\hat{\tau}^2$	logL
			Model with linear trend				
$\kappa = 0.5$	919.1	−5.58	−15.52	1731.8	2.49	0	−242.71
$\kappa = 1.5$	912.49	−4.99	−16.46	1693.1	0.81	34.9	−240.08
$\kappa = 2.5$	912.14	−4.81	−17.11	1595.1	0.54	54.72	−239.75

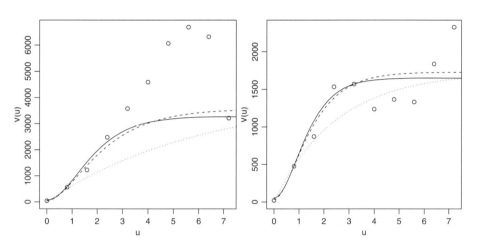

Figure 5.8. Comparison between the sample variogram of the elevation data and Matérn models fitted by maximum likelihood. The solid line corresponds to fits using $\kappa = 2.5$, the dashed line to $\kappa = 1.5$ and the dotted line to $\kappa = 0.5$. Circles correspond to the sample variogram. The left-hand panel shows fits for the model with constant mean, the right-hand panel for the model including a linear trend surface.

terion automatically takes account of the fact that sample variogram ordinates become less precise as u increases, and discounts their influence accordingly. With $\kappa = 1.5$, the maximum likelihood estimates of the remaining covariance parameters are $\hat{\tau}^2 = 48.16$, $\hat{\sigma}^2 = 3510.1$, and $\hat{\phi} = 1.2$ whilst the maximum likelihood estimate of the constant mean is $\hat{\mu} = 848.32$. Note that $\hat{\tau}^2$ is very much smaller than $\hat{\sigma}^2$, consistent with our intuition that surface elevations can be measured with relatively small error, whilst the value of $\hat{\phi}$ indicates that the practical range is approximately $u = 5.7$, and that the spatial correlation has decayed essentially to zero at distances greater than about 8 units. The exact value of the fitted correlation at $u = 8$ is 0.01.

We now re-fit the model, but including a linear trend surface to describe a spatially varying mean, $\mu(x)$. The lower half of Table 5.1 gives the parameter estimates and maximised log-likelihoods. The right-hand panel of Figure 5.8 compares the three fitted variograms with the sample variogram of the residuals from the fitted trend surface; note that the sample variogram shown here differs somewhat from the sample variogram shown previously as Figure 5.5, which was based on ordinary least squares residuals from a quadratic trend surface.

Both inspection of the fitted variograms and comparison of maximised log-likelihoods again lead to the conclusion that the fits for $\kappa = 1.5$ and 2.5 are similar to each other, and qualitatively better than the fit obtained with $\kappa = 0.5$. We choose $\kappa = 1.5$ to enable a direct comparison with the results obtained under the assumption of a constant mean, although the likelihood criterion marginally favours $\kappa = 2.5$. Maximum likelihood estimates of the covariance parameters when $\kappa = 1.5$ are $\hat{\tau}^2 = 34.9$, $\hat{\sigma}^2 = 1693.13$, and $\hat{\phi} = 0.81$. The most striking difference between these results and those obtained under the assumption of a constant mean is the large reduction in the estimate of σ^2. This arises because the trend surface is able to explain a substantial proportion of the spatial variation in observed elevations.

The estimated nugget variance, $\hat{\tau}^2$, is again very small by comparison with $\hat{\sigma}^2$. Finally, the estimate $\hat{\phi}$ indicates that the practical range is now approximately 3.8 and the spatial correlation has decayed essentially to zero by a distance of approximately 5 units (the exact value of the fitted correlation at $u = 5$ is 0.015), somewhat less than for the analysis under the assumption of a constant mean, and again reflecting the fact that the trend surface now accounts for a substantial proportion of the spatial variation in the data.

5.4.5   Restricted maximum likelihood estimation for the Gaussian linear model

A popular variant of maximum likelihood estimation is *restricted maximum likelihood estimation*, or REML. This method of estimation was introduced by Patterson and Thompson (1971) in the context of variance components estimation in designed experiments, for example in animal breeding experiments where the goal is to partition the total variation in a quantitative trait of interest into its genetic and environmental components. In this setting, the small-sample bias of maximum likelihood estimation can be substantial.

Under the assumed model for $\mathrm{E}[Y] = D\beta$, we can transform the data linearly to $Y^* = AY$ such that the distribution of Y^* does not depend on β. Then, the REML principle is to estimate the parameters $\theta = (\nu^2, \sigma^2, \phi)$, which determine the covariance structure of the data, by maximum likelihood applied to the transformed data Y^*. We can always find a suitable matrix A without knowing the true values of β or θ. For example, the projection to ordinary least squares residuals,

$$A = I - D(D^{\mathrm{T}}D)^{-1}D^{\mathrm{T}},$$

has the required property.

Because Y^* is a linear transformation of Y, it retains a multivariate Gaussian distribution. However, the constraint imposed by the requirement that the distribution of Y^* must not depend on β reduces the effective dimensionality of Y^* from n to $n - p$, where p is the number of elements of β.

The REML estimator for θ is computed by maximising the profile likelihood for θ based on the transformed data Y^*. In fact, this can be written in terms of the original data Y as

$$L^*(\theta) = -0.5\{n\log(2\pi) + \log|\sigma^2 V| + \log|D^{\mathrm{T}}\{\sigma^2 V\}^{-1}D|$$
$$+(y - D\tilde{\beta})^{\mathrm{T}}\{\sigma^2 V\}^{-1}(y - D\tilde{\beta})\},$$

where $\sigma^2 V$ is the variance matrix of Y and $\tilde{\beta} = \hat{\beta}(V)$ denotes the maximum likelihood estimator for β for a given value of θ, as given by (5.14). Note that the expression for $L^*(\theta)$ includes an extra determinant term by comparison with the ordinary log-likelihood given by (5.13), and that the matrix A does not appear explicitly i.e., the REML estimate does not depend on the choice of A. The explanation for this is that the condition on A requires it to define a projection of y onto the sub-space of dimension $n-p$ orthogonal to the sub-space of dimension p spanned by the assumed model for the mean response. Different choices of A then correspond to different coordinate systems within the same sub-space, and maximum likelihood estimation is invariant with respect to the choice of coordinates.

Some early references to REML estimation in the geostatistical context are Kitanidis (1983) and Zimmerman (1989). In general, REML leads to less biased estimators for variance parameters in small samples. For example, the elementary unbiased sample variance, $s^2 = (n - 1)^{-1}\sum_{i=1}^{n}(y_i - \bar{y})^2$, is the REML estimator for the variance in a model with constant mean and independent residuals. Note that $L^*(\theta)$ depends on D, and therefore on a correct specification of the model for $\mu(x)$. For designed experiments, the specification of the mean $\mu(x)$ is usually not problematic. However, in the geostatistical setting the specification of the mean $\mu(x)$ is often a pragmatic choice. Although REML is widely recommended for geostatistical models, our experience has been that it is more sensitive than ML to the chosen model for $\mu(x)$.

Harville (1974) showed that REML estimation can also be given a Bayesian interpretation, in the sense that projection of the data onto the residual space is equivalent to ignoring prior information about the mean parameters, β, when making inferences about the covariance parameters, θ.

5.4.6 Trans-Gaussian models

We now consider the transformed Gaussian model in which the transformation is chosen within the Box-Cox family (3.12). We use the word "chosen" rather than "estimated" because of the special role played by the transformation parameter, and the fact that in practice we do not necessarily use formal inferential methods to select a particular transformation. In this section, we denote by $Y = (Y_1, \ldots, Y_n)$ the original response vector, and by $Y^* = (Y_1^*, \ldots, Y_n^*)$ the transformed response. The expectation of Y^* is specified by a linear model, $\mu = D\beta$, whilst the variance matrix of Y^* is written as

$\sigma^2 V(\phi, \nu^2) = \sigma^2 \{R(\phi) + \nu^2 I\}$, where $\nu^2 = \tau^2/\sigma^2$ is the noise-to-signal variance ratio. The log-likelihood including the transformation parameter is

$$
\begin{aligned}
L(\beta, \sigma^2, \phi, \nu^2, \lambda) &= (\lambda - 1) \sum_{i=1}^{n} \log y_i - 0.5\{n \log(2\pi) + \log |\sigma^2 V(\phi, \nu^2)|\} \\
&\quad + (y^* - D\beta)^{\mathrm{T}} \{\sigma^2 V(\phi, \nu^2)\}^{-1} (y^* - D\beta)\}, \quad (5.17)
\end{aligned}
$$

in which the first term arises from the Jacobian of the transformation. Note that (5.17) breaks down if any of the y_i are less than or equal to zero. If zeros occur only because small positive values are rounded down, a simple solution is to impute non-zero values within the rounding range. If genuine zeros are a feature of the data, in the sense that the distribution of Y has a probability mass at zero, the model is strictly inappropriate.

As described in Section 5.4.2, we can obtain explicit estimators for β and σ^2 given ϕ, ν^2 and λ. Full maximum likelihood estimation then requires numerical maximisation with respect to ϕ, ν^2 and λ jointly. Our preferred method of implementation is first to examine the profile log-likelihood for λ, maximising with respect to all remaining model parameters, and to choose an estimate $\tilde{\lambda}$ from amongst a small number of readily interpretable values. These would usually include $\lambda = 1$ (no transformation), $\lambda = 0.5$ (square-root), $\lambda = 0$ (logarithm) and $\lambda = -1$ (reciprocal), but might extend to other rational fractions. If the sole aim of the analysis is empirical prediction of an underlying continuous spatial surface, we might allow any real value of λ. However, our experience is that in most applications, this can be a serious impediment to the scientific interpretation, and hence acceptability, of the statistical analysis. Note also that the profile likelihood can be used to construct an approximate confidence interval for λ, by collecting all values of λ whose associated log-profile-likelihoods lie within one-half the corresponding critical value of the χ_1^2 distribution. Our experience has been that only rarely will the resulting interval unequivocally exclude all of our "readily interpretable" values of λ.

Constructing the profile likelihood for λ is computationally demanding for large data-sets. However, most of the information about λ derives from the marginal distribution of the response variable. Because of this, a simple and effective strategy if we require only a point estimate of λ is to maximise the likelihood under the (false) assumption that the Y_i are mutually independent,

$$
L_0(\beta, \sigma^2, \lambda) = (\lambda - 1) \sum_{i=1}^{n} \log y_i - 0.5\{n \log(2\pi) + n \log |\sigma^2| - \sum_{i=1}^{n} (y_i^* - \beta)^2/\sigma^2\}.
$$

5.4.7 Analysis of Swiss rainfall data

Figure 5.9 shows 467 locations in Switzerland where daily rainfall measurements are taken on 8 May 1986. The resulting data-set of rainfall measurements for 8 May 1986 was used in the project *Spatial Interpolation Comparison 97*; see Dubois (1998) for a detailed description of the data and project. Observed

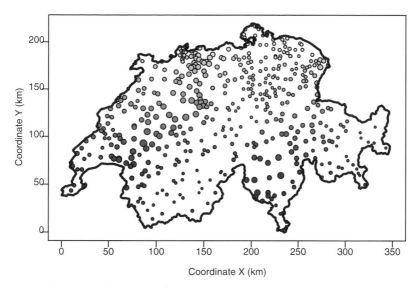

Figure 5.9. Swiss rainfall data. Sampling locations are shown as circles, with the radius of each circle proportional to the corresponding measured rainfall on 8 May 1986. Distances are in kilometres.

rainfall values y_i are recorded as integers, where the unit of measurement is $1/10$ mm. There are five locations where the observed value is equal to zero.

A physically natural model for rainfall would need to take account of known, large-scale meteorological effects, and to include a binary process to model whether or not there is rain, together with a positive-valued process to model the level of rain conditional on it being non-zero. Our purpose here is primarily to illustrate the implementation of the transformed Gaussian model using a well-known data-set. For this reason, we adopt the pragmatic strategy of replacing each zero by the value 0.5. Because only 5 out of 467 responses are affected, the practical effect of this is small, and using other imputed values smaller than 1 had a negligible impact on the results. We therefore assume that the observed rainfall levels, $y = (y_1, \ldots, y_{467})$, form a realisation of the transformed Gaussian model with Matérn correlation function, transformation parameter λ to be chosen within the Box-Cox class (3.12), and a constant mean response, μ.

As described above, we first focus on the estimation of the transformation parameter λ. When the Matérn shape parameter κ is fixed at each of the values $\kappa = 0.5, 1$ and 2, the maximum likelihood estimates of λ are $\hat{\lambda} = 0.514$, 0.508 and 0.508, respectively. The corresponding maximised values of the log-likelihood are -2464.25, -2462.41 and -2464.16. As anticipated, $\hat{\lambda}$ shows very little change in response to changes in κ. For comparison, the estimate of λ obtained by maximising the simpler criterion (5.18), in which we ignore the spatial correlation structure of the model, is $\tilde{\lambda} = 0.537$.

Figure 5.10 shows the profile log-likelihood for λ, holding κ fixed at each of $\kappa = 0.5, 1$ and 2 and maximising with respect to the remaining model parameters, σ^2, ϕ and ν^2. In each case, neither the un-transformed ($\lambda = 1$) nor the log-transformed ($\lambda = 0$) model fits the data well, whereas a square-root trans-

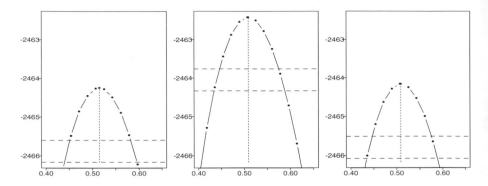

Figure 5.10. Swiss rainfall data. Profile likelihoods for λ, holding the Matérn shape parameter κ fixed. The left panel has $\kappa = 0.5$, the middle panel $\kappa = 1$, the right panel $\kappa = 2$. The two horizontal lines on each plot define approximate 90% and 95% confidence intervals for λ, based on the asymptotic $\frac{1}{2}\chi^2(1)$-distribution of the log-likelihood ratio.

Table 5.2. Swiss rainfall data. Maximum likelihood estimates and maximised values of the log-likelihood, holding the transformation parameter fixed at $\lambda = 0.5$ and the Matérn shape parameter κ taking values $\kappa = 0.5$, 1.0 and 2.0.

κ	$\hat{\mu}$	$\hat{\sigma}^2$	$\hat{\phi}$	$\hat{\tau}^2$	$\log \hat{L}$
0.5	18.36	118.82	87.97	2.48	-2464.315
1	20.13	105.06	35.79	6.92	-2462.438
2	21.36	88.58	17.73	8.72	-2464.185

formation ($\lambda = 0.5$) almost maximises the likelihood, and lies well within a likelihood-based 90% confidence interval for λ. We therefore perform maximum likelihood estimation for the remaining model parameters with λ held fixed at 0.5. The resulting estimates are shown in Table 5.2, together with corresponding values of the maximised log-likelihood.

The final column of Table 5.2 shows that $\kappa = 1$ gives a slightly better fit to the data than either $\kappa = 0.5$ or $\kappa = 2$; the differences between log-likelihoods are 1.87 and 1.75 respectively, both of which lie between the 5% and 10% critical values for a likelihood ratio test.

Table 5.2 also shows that $\hat{\tau}^2$ increases with κ, because an increase in the assumed smoothness of the Gaussian field (as measured by its mean-square differentiability) is compensated by a corresponding increase in the estimated nugget variance, $\hat{\tau}^2$. Notice also the non-orthogonality between κ and ϕ; as κ increases, $\hat{\phi}$ decreases. This again illustrates a general feature of the Matérn model, namely that interpretation of ϕ cannot be made independently of κ.

Figure 5.11 shows the profile log-likelihoods for each of the parameters σ^2, ϕ and $\tau^2 = \sigma^2\nu^2$ when $\kappa = 1$. The profiles indicate the considerable uncertainty with which these parameters, but in particular σ^2 and ϕ, are estimated despite the relatively large size of the data-set. Note also that the profile log-likelihoods for σ^2 and ϕ are clearly asymmetric, suggesting that their sampling distributions may be markedly non-Gaussian.

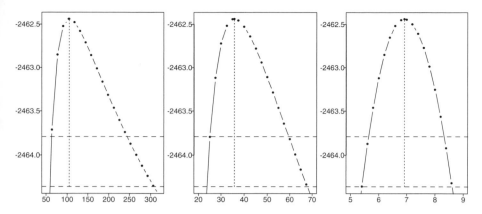

Figure 5.11. Swiss rainfall data. Profile log-likelihoods for covariance parameters σ^2 (left panel), ϕ (middle panel) and τ^2 (right panel), when $\kappa = 1$ and $\lambda = 0.5$. The two horizontal lines on each plot define approximate 90% and 95% percent confidence intervals for λ, based on the asymptotic $\frac{1}{2}\chi^2(1)$-distribution of the log-likelihood ratio.

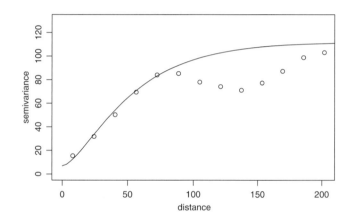

Figure 5.12. Swiss rainfall data. Sample variogram of square-root transformed data (dot-dashed line), compared with fitted theoretical variogram (solid line).

Figure 5.12 compares the sample semivariogram of the square-root trans-formed data with the fitted theoretical variogram. The fit is good, at least for small distances where the sample variogram has relatively high precision. This again illustrates how the likelihood discounts the potentially wayward influence of empirical variogram ordinates at large inter-point distances.

5.4.8 Analysis of soil calcium data

Consider now the calcium content variable in the soil data-set described in Example 1.4. Contextual information on soil usage, together with exploratory analysis of these data, suggests the need for a spatially varying model for the mean, which we now investigate further. The potential covariates are: the soil

Table 5.3. Number of parameters and maximised log-likelihoods for the models fitted to the Calcium data-set.

Model	Parameters	2 logL
M1	4	-1265.36
M2	6	-1258.65
M3	7	-1258.12
M4	8	-1255.56
M5	9	-1255.46

type (as delineated by the three sub-areas within the study region); elevation; and the spatial coordinates themselves. We codify different model-specifications for the mean as follows:

M1: constant

M2: soil type

M3: soil type and elevation

M4: soil type and a linear trend on the coordinates

M5: soil type, elevation and a linear trend on the coordinates

We assume a Matérn model with $\kappa = 0.5$ (exponential) for the correlation function. Models with $\kappa = 1.5$ or 2.5 did not improve the fit of the model to the data. Table 5.3 gives the maximised log-likelihoods of the five candidate models for the mean response, which have 3 covariance parameters (σ^2, τ^2, ϕ) and 1, 3, 4, 5 and 6 mean parameters, respectively.

We first included soil type in the model because previous managements practices are expected to have a direct effect on the calcium content.

For model choice we can use the log-likelihood-ratio criterion to compare nested models. For these data, in accordance with the results of our exploratory analysis, the log-likelihood-ratio criterion favours a model with different means for the three sub-areas. Neither elevation nor a linear trend surface gave significant improvements in the maximised log-likelihood.

For the chosen model, the estimates of the mean parameters are $\hat{\beta} = (39.71, 47.75, 53.52)$, corresponding to the estimated mean values in each of the three sub-areas. The estimates of the three covariance parameters are $\hat{\sigma}^2 = 98.7$, $\hat{\phi} = 72.61$ and $\hat{\tau}^2 = 3.26$.

Figure 5.13 illustrates a variogram-based diagnostic for the fitted model. It shows the variogram obtained using the estimated stochastic components of the fitted model i.e., $Y - D\hat{\beta}$ where Y is the measured calcium, D is the matrix whose columns consist of dummy variables identifying the three sub-areas and $\hat{\beta}$, as above, is the estimated mean for each of the three sub-areas. This variogram is compared with simulation envelopes obtained by repeatedly simulating from the fitted model at the data locations. Notice the asymmetry of the envelope relative to the data-based variogram and, more particularly, the width of the envelope. This example underlines the difficulty of discriminating empirically between different members of the Matérn family for data-sets of this size.

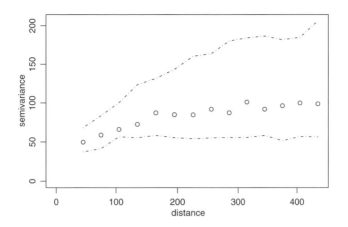

Figure 5.13. Diagnostic for the soil data with the sample variogram and envelopes obtained by simulating from the fitted model.

5.5 Parameter estimation for generalized linear geostatistical models

The application of likelihood-based methods to non-Gaussian generalized linear geostatistical models is hampered by computational difficulties, which arise because of the high dimensionality of the unobserved random vector $S = \{S(x_1), \ldots, S(x_n)\}$.

In a generalized linear mixed model, the likelihood function has a simple, explicit form conditional on the values of a vector of unobserved random variables, S, which are usually called random effects. This simplicity is a consequence of the key assumption that the responses, $Y = (Y_1, \ldots, Y_n)$, are conditionally independent given S. Let θ denote parameters which determine the conditional distribution of Y given S, and write $f_i(y_i; S, \theta)$ for the conditional distribution of Y_i given S and θ. Then, the conditional likelihood for θ were S to be observed is

$$L(\theta|S) = \prod_{i=1}^{n} f_i(y_i|S, \theta). \tag{5.18}$$

Now, let $g(S; \phi)$ denote the joint distribution of S, with parameter ϕ. Then, from a classical perspective the likelihood function based on the observed random variables Y is obtained by marginalising with respect to the unobserved random variables S, leading to the mixed-model likelihood,

$$L(\theta, \phi) = \int_S \prod_{i=1}^{n} f_i(y_i|S, \theta) g(s|\phi) ds. \tag{5.19}$$

If the S_i are mutually independent, the multiple integral in (5.19) reduces to a product of one-dimensional integrals, and numerical evaluation of the mixed-model likelihood is relatively straightforward. The difficulty in applying (5.19) in the geostatistical setting is that the $S_i = S(x_i)$ are dependent, the integral in (5.19) therefore has the same dimension as Y and conventional methods of numerical integration fail. Breslow and Clayton (1993) used approximate methods of integration. However, the accuracy of these approximate methods in high-dimensional problems is unclear, and they are especially problematic when, as is typical of geostatistical problems, the variability in the distribution of S is large relative to the variability in the conditional distribution of Y given S.

5.5.1 Monte Carlo maximum likelihood

Developments in Monte Carlo methods, including key contributions by Geyer and Thompson (1992) and Geyer (1994), provide ways to construct better approximations to the log-likelihood function of generalised linear mixed models (GLMM's). Zhang (2002) develops a Monte Carlo version of the EM algorithm for maximum likelihood estimation of parameters of generalised linear geostatistical models (GLGM's), using a Metropolis-Hastings algorithm to produce samples of the random effects at the sample sites. Christensen (2004) describes a more general approach, constructing Monte Carlo approximations to the likelihood or profile likelihood functions by means of an MCMC algorithm for simulating from the conditional distribution of the random effects.

The integral which defines the likelihood function (5.19) can be expressed as an expectation with respect to the distribution of S, namely

$$L(\theta, \phi) = \mathrm{E}\left[\prod_{i=1}^{n} f_i(y_i|S, \theta)\right]. \tag{5.20}$$

Hence, in principle, for any set of values of (θ, ϕ) we can simulate repeatedly from the corresponding multivariate Gaussian distribution of S and approximate the expectation by a Monte Carlo average,

$$L_{MC}(\theta, \phi) = K^{-1} \sum_{k=1}^{K}\left[\prod_{i=1}^{n} f_i(y_i|S_k, \theta)\right], \tag{5.21}$$

where S_k denotes the kth simulated realisation of the vector S. In practice, using an independent random sample of S_k to evaluate (5.21) is likely to prove hopelessly inefficient, and some kind of variance reduction technique is needed. Chapter 5 of Ripley (1987) gives a general discussion of variance reduction techniques. Section 3.15.1 of Geyer (1994) shows how Markov chain Monte Carlo can be used, treating S as missing data.

Despite these advances in Monte Carlo methods, there is still a place for computationally simpler approaches which can be used routinely, especially when a range of candidate models are under consideration. We therefore describe briefly two alternative approaches which have been proposed for parameter estimation within generalized linear mixed models.

5.5.2   Hierarchical likelihood

Lee and Nelder (1996) propose an unconventional definition of the likelihood function for generalized linear mixed models which they call hierarchical likelihood. Using the same model and notation as in (5.19), their hierarchical log-likelihood function is

$$L(\theta, \phi) = \sum_{i=1}^{n} \log f_i(y_i|S, \theta) + \log g(s|\phi). \tag{5.22}$$

Point estimates of θ, ϕ and of the unobserved values of S, are then obtained by maximisation of 5.22. This is equivalent to a form of penalised log-likelihood in which different values of S are penalized according to the likelihood of their occurrence under the assumed distribution for S. Maximisation of the hierarchical likelihood avoids the need to integrate with respect to the distribution of S. Although the published discussion of Lee and Nelder (1996) raised doubts about the properties of the associated inferences in the high-dimensional case which always applies in the geostatistical setting, Lee and Nelder (2001) show how the method can be used in conjunction with careful diagnostic checking to identify and fit a wide range of models for spatially or temporally correlated data.

5.5.3   Generalized estimating equations

One way round the computational difficulties discussed above is to abandon likelihood-based methods in favour of possibly less efficient but computationally simpler methods. An example, originally proposed by Liang and Zeger (1986) as a strategy for analysing correlated longitudinal data, is the method of generalized estimating equations (GEE). This builds on the idea of quasi-likelihood estimation for the classical generalized linear model, as proposed by Wedderburn (1974). Recall from Section 4.1 that in the classical generalized linear model responses $Y_i : i = 1, \ldots, n$ are assumed to be independent with expectations $\mu_i(\beta)$ specified as known functions of a set of regression parameters β, and the distribution of Y_i is of a known form, but parameterised by μ_i. In a quasi-likelihood model, the specification of the distribution of each Y_i is relaxed to a specification of its variance as a function of its mean, hence $\text{Var}(Y_i) = v(\mu_i)$, where $v(\cdot)$ is known up to a constant of proportionality. Wedderburn (1974) showed that β could then be estimated consistently by solving the *estimating equations*

$$\frac{\partial \mu}{\partial \beta} V^{-1}(Y - \mu) = 0 \tag{5.23}$$

where $Y - \mu$ is the vector with elements $Y_i - \mu_i(\beta)$ and V is the diagonal matrix with diagonal elements $v(\mu_i)$. The resulting estimates have similar properties to maximum likelihood estimates in a fully specified probability model, indeed they are the maximum likelihood estimates in an exponential family probability model with the stated mean and variance structure. The "generalization" of (5.23) which leads to GEE is to allow a non-diagonal V, so as to reflect correlations amongst the Y_i.

The GEE method was devised by Liang and Zeger to solve problems of the following kind. The data consist of many independent replications of a low-dimensional vector of responses, Y say, and the scientific objective concerns inference about the unconditional mean response vector, $E(Y)$. A typical application would be to longitudinal studies in public health, where the independent replication arises from different subjects in the study, and the required inferences concern the effect of a treatment intervention on the longitudinal population mean response. Most geostatistical applications are not of this kind. Geostatistical data typically consist of a single realisation of an n-dimensional response Y, and the questions of scientific interest are often concerned more with spatial prediction than with inference about $E(Y)$.

Nevertheless, Gotway and Stroup (1997) develop a version of GEE for the geostatistical data. Their approach to model-fitting consists of first fitting a classical generalized linear model to the data i.e., temporarily ignoring any spatial dependence in the data, then using the empirical variogram of the standardised residuals from this preliminary fit to identify a model for the spatial correlation. The model can then be re-fitted by solving the estimating equations 5.23, incorporating the identified model for the spatial correlation into the specification of the variance matrix V. Note that, as in more traditional applications of the GEE approach, the parameters β do not have the same meaning as they do in the generalized linear mixed model; see, for example, the discussion in Chapter 7 of Diggle, Heagerty, Liang and Zeger (2002). With this proviso, Gotway and Stroup's approach is appealing when the scientific focus is on the way in which explanatory variables affect the mean response. In these circumstances, GEE gives a simple way of making the required inferences whilst adjusting for spatial correlation. GEE is, in our opinion, less attractive when the scientific focus is on estimation of spatial dependence, or on spatial prediction, because of the somewhat *ad hoc* way in which the spatial dependence parameters are estimated. In fact, the prediction equations proposed in Gotway and Stroup (1997) are equivalent, in our terms, to plug-in prediction under a Gaussian model, as set out in Section 2.6, but with a non-linear regression model for the mean response.

5.6   Computation

5.6.1   Variogram calculations

We now give examples of variogram calculations, using the function `variog()` in conjunction with both real and simulated data. We start by showing the commands used to produce Figures 5.5 and 5.2.

```
> data(elevation)
> plot(variog(elevation, option = "cloud"), xlab = "u",
+      ylab = "V(u)")
```

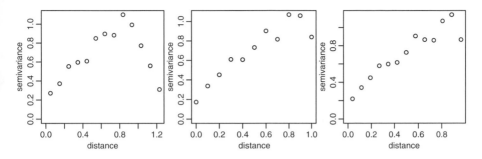

Figure 5.14. Variograms for the surface elevation data with different options for the bin size.

```
> plot(variog(elevation, uvec = seq(0, 8, by = 0.5)),
+       xlab = "u", ylab = "V(u)")
> plot(variog(elevation, trend = "2nd", max.dist = 6.5),
+       xlab = "u", ylab = "V(u)")
```

By experimenting with different bin specifications for the sample variogram of a single data-set, the reader can gain useful insight into the extent to which the resulting variogram estimates are sensitive to this choice. To illustrate this we use the simulated data-set s100, which is included in **geoR** and is accessed by the following command.

```
> data(s100)
```

These data are simulated from a model with a constant mean equal to zero, unit signal variance, zero nugget variance and exponential correlation function with $\phi = 0.3$. The commands below produce Figure 5.14 which shows sample variograms obtained with different binning options.

```
> v1 <- variog(s100)
> plot(v1)
> v2 <- variog(s100, uvec = seq(0, 1, by = 0.1))
> plot(v2)
> v3 <- variog(s100, max.dist = 1)
> plot(v3)
```

Notice that the binning can be user-defined via the uvec argument, or by the max.dist argument. By default, 13 bins are defined spanning the range of distances from zero to the maximum distance between any two pairs of data locations. As discussed earlier in this chapter, extending the variogram calculations over the full available range of inter-point distances is not necessarily helpful, because the variogram estimates often become unstable at large distances. The midpoints of the bins obtained from the commands listed above are:

```
> round(v1$u, dig = 2)

 [1] 0.05 0.15 0.25 0.34 0.44 0.54 0.64 0.74 0.84 0.93 1.03 1.13
[13] 1.23
```

```
> round(v2$u, dig = 2)
 [1] 0.0 0.1 0.2 0.3 0.4 0.5 0.6 0.7 0.8 0.9 1.0
> round(v3$u, dig = 2)
 [1] 0.04 0.12 0.19 0.27 0.35 0.42 0.50 0.58 0.65 0.73 0.81 0.88
[13] 0.96
```

The three panels of Figure 5.14 show qualitatively similar patterns at small distances but clear quantitative differences which would be reflected in the results of any variogram-based parameter estimation method. Note that we have not attempted to impose common x and y axis scales on the three panels of Figure 5.14, although this could be done using additional, optional arguments to the variog() function.

The soil data of Example 1.4 include two response variables, calcium and magnesium content, and two potential explanatory variables, area and altitude, which identify the sub-area (or soil type) and elevation, respectively, for each data location. An exploratory plot produced with plot(ca20, trend= area+altitude, low=T) suggests a possible quadratic trend. Here, we use the calcium content response to construct the four variograms shown in Figure 5.15. The upper-left panel shows a variogram calculated from the original data, whilst the upper-right panel uses residuals from a linear model adjusting for sub-region as a factor on three levels. The lower-left panel uses residuals adjusting for both sub-region and elevation. Finally, the lower-right panel adjusts for sub-region, elevation and a quadratic trend surface. The differences amongst the resulting variograms again illustrate the inter-play between the specification of a model for the mean response and the resulting estimated covariance structure. The results in this case indicate that by incorporating sub-region into a model for the calcium content we appear to achieve approximate stationarity of the residuals, because the corresponding variogram reaches a plateau. On the evidence of Figure 5.15 alone, it is less clear whether the trend surface is needed; if it is included in the model, its effect is to reduce the effective range of the spatial correlation. The code to produce Figure 5.15 is as follows.

```
> data(ca20)
> plot(variog(ca20, max.dist = 510))
> plot(variog(ca20, trend = ~area, max.dist = 510))
> plot(variog(ca20, trend = ~area + altitude, max.dist = 510))
> t.all <- trend.spatial(ca20, trend = ~area + altitude,
+       add = "2nd")
> plot(variog(ca20, trend = ~t.all, max.dist = 510))
```

Applying the variog() function to replicated simulations of a Gaussian process can help to understand the typical pattern of variation which the sample variogram exhibits relative to the underlying theoretical variogram. Each panel of Figure 5.16 shows the true variogram as a smooth curve, together with sample variograms of three simulations of the process on 100 locations in a unit square. To generate Figure 5.16 we simulated from the Gaussian model with Matérn correlation function and parameters $\kappa = 1.5$, $\sigma^2 = 1$, $\tau^2 = 1$ in both

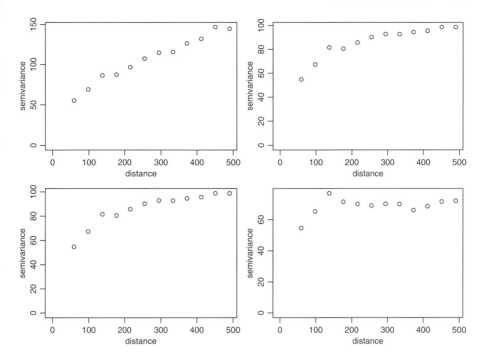

Figure 5.15. Sample variograms for the `ca20` data. The upper-left panel uses the unadjusted calcium response, the upper-right uses the residuals after adjusting for sub-region, the lower-left panel uses residuals after adjusting for sub-region and elevation, the lower-right panel uses residuals after adjusting for sub-region, elevation and a quadratic trend surface.

cases. For the left-hand panel we set $\phi = 0.05$, and for the right-hand panel $\phi = 0.2$. Note the substantial variation amongst the three sampled variograms within each panel. The code follows.

```
> set.seed(83)
> sim1 <- grf(100, cov.pars = c(1, 0.05), cov.model = "mat",
+     kap = 1.5, nsim = 3)
> plot(variog(sim1, max.dist = 1), type = "l", lty = 1:3,
+     col = 1)
> lines.variomodel(seq(0, 1, l = 100), cov.model = "mat",
+     kap = 1.5, cov.pars = c(1, 0.05), nug = 0)
> set.seed(83)
> sim2 <- grf(100, cov.pars = c(1, 0.2), cov.model = "mat",
+     kap = 1.5, nsim = 3)
> plot(variog(sim2, max.dist = 1), type = "l", lty = 1:3,
+     col = 1)
> lines.variomodel(seq(0, 1, l = 100), cov.model = "mat",
+     kap = 1.5, cov.pars = c(1, 0.2), nug = 0)
```

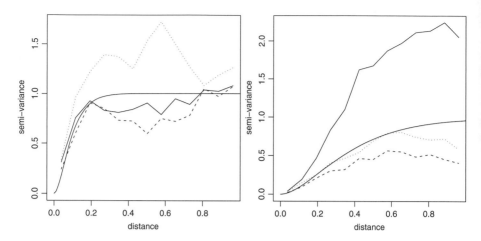

Figure 5.16. Sample variograms for three simulations of a process with Matérn correlation function. Parameter values are $\kappa = 1.5$, $\sigma^2 = 1$, $\tau^2 = 0$ in both panels. In the left-hand panel, $\phi = 0.05$, in the right-hand panel $\phi = 0.2$. The smooth curve in each panel is the theoretical variogram.

5.6.2 Parameter estimation

The next examples show how **geoR** can be used to estimate parameters in Gaussian models, with a particular focus on parameters which define the covariance structure of the model. We illustrate two different approaches, *ad hoc* curve-fitting methods and maximum likelihood, again using the simulated data s100 included with the package.

Parameter estimation using a variogram-based method is done in two steps. In the first step, we calculate and plot the sample variogram, and experiment with "fitted by-eye" parameter values, as in the following code which produces Figure 5.17. With **geoR**, this can be achieved either by passing model information to the lines.variomodel() function, or by using the interactive function eyefit().

```
> s100.v <- variog(s100, max.dist = 1)
> plot(s100.v)
> lines.variomodel(seq(0, 1, l = 100), cov.pars = c(0.9,
+      0.2), cov.model = "mat", kap = 1.5, nug = 0.2)
```

In the second step, we use the function variofit() to implement a curve-fitting method. The function takes as argument the sample variogram. Optional arguments allow the user to specify different types of weights for the least-squares criterion, with n-weighted least squares as the default. The function calls a numerical minimisation algorithm and therefore also requires initial values for the parameters. Here, we use the ones previously fitted by eye. The correlation function of choice can also be specified, with the exponential model as the default. Also by default, the κ parameter for the Matérn or powered exponential family is held fixed and the estimation is restricted to the parameters σ^2, τ^2 and ϕ. The following code illustrates the process.

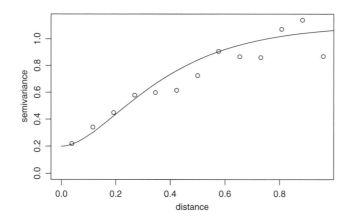

Figure 5.17. Empirical variogram for the s100 data-set (circles) and a line indicating a model fitted by eye.

```
> wls <- variofit(s100.v, ini = c(0.9, 0.2), cov.model = "mat",
+     kap = 1.5, nug = 0.2)
> wls

variofit: model parameters estimated by WLS (weighted least squares):
covariance model is: matern with fixed kappa = 1.5
parameter estimates:
  tausq sigmasq     phi
 0.3036  0.9000  0.2942

variofit: minimised weighted sum of squares = 23.6572
```

Estimation by the maximum likelihood method is implemented in the function likfit(). The sample variogram is strictly not required for the maximum likelihood method, but nevertheless provides a useful way to specify initial values for the numerical minimisation algorithm.

The input to likfit() specifies the data object, the model choice and initial values. As with the variofit() function, the value of κ is held fixed by default, although it can also be estimated by setting the optional argument fix.kappa=FALSE. The nugget parameter is included in the estimation by default, but can be held fixed using the optional argument fix.nugget=TRUE. The argument method allows for maximum likelihood or restricted maximum likelihood. An example follows.

```
> ml <- likfit(s100, cov.model = "mat", kap = 1.5, ini = c(0.9,
+     0.2), nug = 0.2)
> ml

likfit: estimated model parameters:
    beta    tausq  sigmasq      phi
"0.8964" "0.0000" "0.7197" "0.0476"
```

```
likfit: maximised log-likelihood = -85.26

> reml <- likfit(s100, cov.model = "mat", kap = 1.5,
+        ini = c(0.9, 0.2), nug = 0.2, met = "reml")
> reml

likfit: estimated model parameters:
    beta    tausq sigmasq      phi
"0.8936" "0.0000" "0.7427" "0.0485"

likfit: maximised log-likelihood = -83.81
```

As our final example we show the commands used to fit the five alternative models to the soil data example presented in Section 5.4.8. The covariates are included by using the argument trend in the call to likfit().

```
> data(ca20)
> m1 <- likfit(ca20, ini = c(100, 200), nug = 50)
> m2 <- likfit(ca20, trend = ~area, ini = c(60, 100),
+        nug = 40)
> m3 <- likfit(ca20, trend = ~area + altitude, ini = c(60,
+        100), nug = 40)
> m4 <- likfit(ca20, trend = ~area + coords, ini = c(60,
+        100), nug = 40)
> m5 <- likfit(ca20, trend = ~area + altitude + coords,
+        ini = c(60, 100), nug = 40)
```

5.7   Exercises

5.1. Fit a quadratic trend surface model to the elevation data, using both maximum likelihood and REML to estimate the model parameters, and compare the results.

5.2. Write a programme to simulate data from a stationary Gaussian model with exponential covariance function, $\rho(u) = \sigma^2 \exp(-u/\phi)$. Apply the method of maximum likelihood estimation to replicate simulations of this model, and investigate the joint sampling distribution of $\hat{\sigma}^2$ and $\hat{\phi}$. Compare this with the joint sampling distribution for Zhang's re-parameterisation to $\theta_1 = \log(\sigma^2/\phi)$ and $\theta_2 = \log(\phi)$.

5.3. Design and implement a study to compare the performance of ordinary (n-weighted) least squares, weighted least squares and Gaussian maximum likelihood estimation for the parameters of a stationary Gaussian model with exponential covariance function.

5.4. Extend your simulation study from Exercise 5.3 to included a model with t-distributed margins in place of the Gaussian. Note that if $Z = (Z_1, \ldots, Z_n)$ is an independent random sample from any distribution with zero mean and unit variance, then $Y = HZ$ has zero mean and covariance matrix HH', irrespective of the marginal distribution of the Z_i.

6
Spatial prediction

In this chapter, we consider the problem of using the available data to predict aspects of the realised, but unobserved, signal $S(\cdot)$. More formally, our target for prediction is the realised value of a random variable $T = \mathcal{T}(S)$, where S denotes the complete set of realised values of $S(x)$ as x varies over the spatial region of interest, A. The simplest example of this general problem is to predict the value of the signal, $T = S(x)$, at an arbitrary location x, using observed data $Y = (Y_1, \ldots, Y_n)$, where each Y_i represents a possibly noisy version of the corresponding $S(x_i)$. Other common targets T include the integral of $S(x)$ over a prescribed sub-region of A or, more challengingly, a non-linear functional such as the maximum of $S(x)$, or the set of locations for which $S(x)$ exceeds some prescribed value. In this chapter, we ignore the problem of parameter estimation, in effect treating all model parameters as known quantities.

6.1 Minimum mean square error prediction

In very general terms, the prediction problem can be stated as follows. Let Y denote a vector of random variables whose realised values are observed, and let T denote any other random variable whose realised value we wish to predict from the observed value of Y. A *point predictor* for T is any function of Y, which we denote by $\hat{T} = t(Y)$.

The *mean square prediction error* of $\hat{T}$ is

$$MSE(\hat{T}) = \mathrm{E}[(\hat{T} - T)^2], \tag{6.1}$$

where the expectation is with respect to the joint distribution of T and $\hat{T}$ or, equivalently, the joint distribution of T and Y. The general form of the point predictor which minimises $MSE(\hat{T})$ is then given by the following well-known result.

Theorem 6.1. $MSE(\hat{T})$ takes its minimum value when $\hat{T} = \mathrm{E}(T|Y)$.
Proof
Write

$$\mathrm{E}[(T - \hat{T})^2] = \mathrm{E}_Y[\mathrm{E}_T[(T - \hat{T})^2|Y]], \qquad (6.2)$$

where the subscripts on the two expectation operators indicate that the expectations are with respect to Y and T, respectively. Write the inner expectation in (6.2) as

$$\mathrm{E}_T[(T - \hat{T})^2|Y] = \mathrm{Var}_T\{(T - \hat{T})|Y\} + \{\mathrm{E}_T[(T - \hat{T})|Y]\}^2.$$

Conditional on Y, any function of Y is a constant, so $\mathrm{Var}_T\{(T - \hat{T})|Y\} = \mathrm{Var}_T(T|Y)$ and $\mathrm{E}_T[T - \hat{T}|Y] = \mathrm{E}_T[T|Y] - \hat{T}$. Hence,

$$\mathrm{E}_T[(T - \hat{T})^2|Y] = \mathrm{Var}_T(T|Y) + \{\mathrm{E}_T(T|Y) - \hat{T}\}^2. \qquad (6.3)$$

Now take the expectation of the expression on the right-hand side of (6.3) with respect to Y. This gives

$$\mathrm{E}[(T - \hat{T})^2] = \mathrm{E}_Y[\mathrm{Var}_T(T|Y)] + \mathrm{E}_Y\{[\mathrm{E}_T(T|Y) - \hat{T}]^2\}. \qquad (6.4)$$

The first term on the right-hand side of (6.4) does not depend on the choice of $\hat{T}$, whilst the second is non-negative, and equal to zero if and only if $\hat{T} = \mathrm{E}(T|Y)$. This completes the proof.

The statement of Theorem 6.1 is strikingly simple and makes intuitive sense. However, it is worth emphasising that it follows from adopting mean square error as the criterion to be optimised, which is not necessarily the most appropriate measure of performance in any specific application. Note in particular that the result is not transformation-invariant i.e., if $\hat{T}$ is the minimum mean square error predictor for T then in general $g(\hat{T})$ is not the minimum mean square error predictor for $g(T)$. A point prediction provides a convenient summary, but a complete answer to the prediction problem is the conditional distribution of T given Y, and the mean of this conditional distribution is simply one of a number of summaries which we could have used.

It follows from (6.4) that the mean square error of $\hat{T}$ is

$$\mathrm{E}[(T - \hat{T})^2] = \mathrm{E}_Y[\mathrm{Var}(T|Y)]. \qquad (6.5)$$

We call $\mathrm{Var}(T|Y)$ the *prediction variance*. The value of the prediction variance at the observed value of Y estimates the achieved mean square error of $\hat{T}$.

Note also that $\mathrm{E}[(T-\hat{T})^2] \leq \mathrm{Var}(T)$, with equality if T and Y are independent random variables. This follows from the fact that $\mathrm{Var}(T) = \mathrm{E}[(T - \mathrm{E}[T])^2]$ is the mean square error of the trivial predictor $\tilde{T} = \mathrm{E}[T]$ which ignores the data Y. Informally, the difference between the marginal variance $\mathrm{Var}(T)$ and the conditional variance $\mathrm{Var}(T|Y)$ gives a summary measure of how useful the data Y are for predicting T.

6.2   Minimum mean square error prediction for the stationary Gaussian model

We now assume that our data $Y = (Y_1, \ldots, Y_n)$ are generated by the stationary Gaussian model as defined in Section 2.2. We write $S = (S(x_1), \ldots, S(x_n))$ for the unobserved values of the signal at the sampling locations $x_1, \ldots, x_n$. Then, S is multivariate Gaussian with mean vector $\mu\mathbf{1}$, where $\mathbf{1}$ denotes a vector each of whose elements is 1, and variance matrix $\sigma^2 R$, where R is the n by n matrix with elements $r_{ij} = \rho(||x_i - x_j||)$. Similarly, Y is multivariate Gaussian with mean vector $\mu\mathbf{1}$ and variance matrix

$$\sigma^2 V = \sigma^2(R + \nu^2 I) = \sigma^2 R + \tau^2 I , \qquad (6.6)$$

where I is the identity matrix.

6.2.1   Prediction of the signal at a point

Suppose initially that our objective is to predict the value of the signal at an arbitrary location, thus our target for prediction is $T = S(x)$. Then, (T, Y) is also multivariate Gaussian and we obtain the minimum mean square error predictor $\hat{T}$ by using the following standard result on the multivariate Gaussian distribution.

Theorem 6.2. Let $X = (X_1, X_2)$ be jointly multivariate Gaussian, with mean vector $\mu = (\mu_1, \mu_2)$ and covariance matrix

$$\Sigma = \left[\begin{array}{cc} \Sigma_{11} & \Sigma_{12} \\ \Sigma_{21} & \Sigma_{22} \end{array} \right],$$

i.e., $X \sim \text{MVN}(\mu, \Sigma)$. Then, the conditional distribution of X_1 given X_2 is also multivariate Gaussian, $X_1|X_2 \sim \text{MVN}(\mu_{1|2}, \Sigma_{1|2})$, where

$$\mu_{1|2} = \mu_1 + \Sigma_{12}\Sigma_{22}^{-1}(X_2 - \mu_2)$$

and

$$\Sigma_{1|2} = \Sigma_{11} - \Sigma_{12}\Sigma_{22}^{-1}\Sigma_{21}.$$

To apply Theorem 6.2 to our prediction problem, note that (T, Y) is multivariate Gaussian with mean vector $\mu\mathbf{1}$ and variance matrix

$$\left[\begin{array}{cc} \sigma^2 & \sigma^2 r' \\ \sigma^2 r & \sigma^2 V \end{array} \right]$$

where r is a vector with elements $r_i = \rho(||x - x_i||) : i = 1, \ldots, n$ and V is given by (6.6).

Then, Theorem 6.2 with $X_1 = T$ and $X_2 = Y$ gives the result that the minimum mean square error predictor for $T = S(x)$ is

$$\hat{T} = \mu + r'V^{-1}(Y - \mu\mathbf{1}) \qquad (6.7)$$

with prediction variance

$$\text{Var}(T|Y) = \sigma^2(1 - r'V^{-1}r). \qquad (6.8)$$

Note that in the special setting of the multivariate Gaussian distribution, the conditional variance does not depend on Y, and the achieved mean square error is therefore equal to the prediction variance.

6.2.2 Simple and ordinary kriging

In traditional geostatistical terminology, construction of the surface $\hat{S}(x)$, where $\hat{T} = \hat{S}(x)$ is given by (6.7), is called *simple kriging*. The name acknowledges the influence of D. G. Krige, who pioneered the use of statistical methods in the South African mining industry (Krige, 1951). Because we are here treating parameters as known, the predictor (6.7) is linear in the data. To use (6.7) in practice, we need to plug-in estimated values for the model parameters.

A common practice in geostatistics is to use a modified kriging algorithm called *ordinary kriging*. The distinction between simple and ordinary kriging is that in the latter, the mean value is treated as unknown, whereas it is still assumed that covariance parameters are known. This leads to a linear predictor similar to (6.7) except that μ is replaced by its generalised least squares estimator,

$$\hat{\mu} = (\mathbf{1}'V^{-1}\mathbf{1})^{-1}\mathbf{1}'V^{-1}Y ,$$

with V given by (6.6). The ordinary kriging predictor can be expressed as a linear combination, $\hat{S}(x) = \sum a_i(x)Y_i$. The $a_i(x)$ are called the *prediction weights*, or *kriging weights*, and have the property that $\sum a_i(x) = 1$ for any target location x.

Some authors reserve the name *simple kriging* to mean (6.7) in conjunction with the plug-in estimate $\hat{\mu} = \bar{y}$, in which case we can again write $\hat{S}(x)$ as a linear combination of the Y_i, but the kriging weights are no longer constrained to sum to one. We give specific examples in Section 6.4. From a model-based perspective, there is no fundamental distinction between simple and ordinary kriging. Both are examples of plug-in prediction; they differ only in respect of which plug-in estimate of μ they use.

In our derivation of simple and ordinary kriging, we begin with a stochastic model for the data, Y, and the signal process $S(x)$, and derive the explicit form of the minimum mean square error predictor for $S(x)$. In this approach, the fact that the predictor is linear in Y is a consequence of the Gaussian modelling assumption. In classical geostatistics, the starting point is to restrict attention to predictors which are linear in Y, and to look for the one which is optimal, in a mean square error sense. The resulting expressions for simple and ordinary kriging predictors can also be derived as examples of the much older statistical ideas of best linear, and best linear unbiased, prediction, respectively. See, for example, chapter 1 of Stein (1999). Ordinary kriging also has a Bayesian interpretation, which we discuss in Chapter 7. Briefly, if we consider all parameters in the stationary Gaussian model to be known except the mean, to which we assign a Gaussian prior distribution, then the resulting posterior mean predictor for $S(x)$ reduces to the ordinary kriging predictor in the limit as the posterior variance tends to infinity. See Kitanidis (1978), Omre (1987), Omre and Halvorsen (1989) and Omre, Halvorsen and Berteig (1989).

6.2.3 Prediction of linear targets

Suppose now that we wish to predict a *linear* target T, by which we mean any target of the form

$$T = \int_A w(x)S(x)dx$$

for some prescribed weighting function $w(x)$. Because expectation is a linear operator, it follows that whatever the model for Y,

$$E[T|Y] = \int_A w(x)E[S(x)|Y]dx, \qquad (6.9)$$

or in other words,

$$\hat{T} = \int_A w(x)\hat{S}(x)dx.$$

Furthermore, under the stationary Gaussian model, (T, Y) is multivariate Gaussian and the predictive distribution of T is univariate Gaussian with mean given by (6.9) and variance

$$\text{Var}(T|Y) = \int_A \int_A w(x)w(x')\text{Cov}\{S(x), S(x')\}dxdx'.$$

In summary, to predict a linear target it is sufficient to predict the values of the signal over the region A of interest and to evaluate the target directly from the predicted surface, $\{\hat{S}(x) : x \in A\}$. This does not apply to non-linear targets. Using non-linear properties of $\hat{S}(x)$ as predictors of the corresponding properties of the true surface $S(x)$ can be a very poor strategy.

6.2.4 Prediction of non-linear targets

The predictive distribution of a non-linear property of the signal is generally intractable, in which case we use a Monte Carlo method based on a conditional simulation of the signal process $S(\cdot)$, given the data Y. In principle, this method solves any non-linear prediction problem associated with the stationary Gaussian model. However, the solution may or may not be computationally feasible in large problems.

We first approximate the continuous region A by a discrete grid of *prediction points*, $x_j^* : j = 1, \ldots, N$ to cover A. We then simulate a realisation of $S^* = \{S(x_j^*) : j = 1, \ldots, N\}$ by sampling from the explicit multivariate Gaussian conditional distribution of S^* given Y, and compute the value, T_1 say, of the target T corresponding to the simulated realisation of S^*. Independent replication of the simulation algorithm s times gives a sequence of realisations of S^* and corresponding values T_k. Then, $T_1, \ldots, T_s$ is an independent random sample of size s from the predictive distribution of T and any property of the empirical distribution of the T_k provides an estimate of the corresponding property of the predictive distribution, with a precision determined by the value of N.

The only non-routine aspect of this procedure is the definition of the prediction grid, which has to be fine enough to give a good approximation to the underlying continuous surface, but not so fine as to make the computations infeasible. Roughly speaking, the stronger the correlation between points a given distance apart, the more coarse the prediction grid can be without serious loss of accuracy, but it is difficult to give explicit rules for general use, because the notion of "fine enough" depends on the character of both the true surface and the required target. A pragmatic strategy is to verify empirically that the finer of two candidate grids does not materially change the predictions of interest. Note that if required, each simulated surface can be generated on a progressively finer sequence of grids, using the result of Theorem 6.2 to fill-in progressively from an initially coarse grid.

6.3 Prediction with a nugget effect

The examples in Section 6.4 below confirm the importance of the parameter τ^2 in determining the properties of the simple kriging predictor. The literal interpretation of this parameter in the stationary Gaussian model is as the conditional variance of an observation, Y_i, given the value of the underlying signal, $S(x_i)$. In practice, as discussed briefly in Section 3.5, τ^2 plays a dual role, accounting for both measurement error and short-range spatial variation as follows. Consider the alternative form of the Gaussian model,

$$Y_i = S(x_i) + Z_i : i = 1, \ldots, n, \tag{6.10}$$

where the Z_i are mutually independent, $N(0, \tau^2)$ random variables. Suppose, as an alternative model, that

$$Y_i = S(x_i) + S^*(x_i) : i = 1, \ldots, n, \tag{6.11}$$

where now $S^*(x)$ is a second stationary Gaussian process, independent of $S(x)$ and with the property that its correlation function $\rho^*(u)$ is zero for all $u \geq u_0$. If no two locations in the sample design are less than distance u_0 apart, then the data will be unable to distinguish between the models (6.10) and (6.11). However, the distinction matters in practice because, if we did believe that model (6.11) was the correct one, we should interpolate the data whereas, under the model (6.10) with $\tau^2 > 0$ we should not.

Traditional geostatistics takes a pragmatic view of this distinction, often fitting model (6.10) but constraining its predictions to interpolate the data. This leads to spikes in the predicted surface $\hat{S}(x)$, the so-called "nugget effect." The traditional name of "nugget variance" for the parameter τ^2 refers indirectly to this pragmatic interpretation, in which an isolated high value, apparently unrelated to values in its close proximity, corresponds to a "nugget" of exceptionally high-grade ore.

A model which includes both interpretations of the nugget effect is

$$Y_i = S(x_i) + S^*(x_i) + Z_i : i = 1, \ldots, n. \tag{6.12}$$

To make this model identifiable, we would need to include coincident locations x_i in the sampling design, as discussed in Section 5.2.3. Under model 6.12, when $x_i = x_j$ the expectation of $\frac{1}{2}(Y_i - Y_j)^2$ is equal to τ^2, the variance of the Z_i. Strictly coincident x_i can sometimes be achieved by sample-splitting prior to measurement. Failing this, including near-coincident pairs of locations in the sampling design is a pragmatic alternative, which we discuss in more detail in Chapter 8.

A final comment is that there may be circumstances in which the target for prediction is $Y(x)$, the prospective measured value at an as-yet unsampled location x, rather than the signal $S(x)$. Under the linear Gaussian model, the point predictions of $Y(x)$ and $S(x)$ would be identical, but prediction intervals would be wider for $Y(x)$ than for $S(x)$ because the corresponding prediction variance includes the nugget effect. Specifically, under the model (6.10), $\text{Var}[Y(x)|Y_1, \ldots, Y_n] = \text{Var}[S(x)|Y_1, \ldots, Y_n] + \tau^2$.

6.4   What does kriging actually do to the data?

In this section, we give several simulated examples to show how the assumed parameter values for the underlying model combine with the data to produce the predicted surface $\hat{S}(x)$, under the assumption that the data are generated by the stationary Gaussian model.

Without any essential loss of generality, we fix the mean and variance of the signal to be $\mu = 0$ and $\sigma^2 = 1$. The nugget variance τ^2 can then be interpreted as a noise-to-signal variance ratio. To complete the specification of the assumed model we need to select a correlation function $\rho(u)$. We shall consider two candidate families: the *exponential* correlation function, $\rho(u) = \exp(-u/\phi)$, and the Matérn correlation function defined by equation (2.1) with $\kappa = 1.5$. These correspond to mean-square continuous and mean-square differentiable processes, respectively. The minimum mean square error predictor for $S(x)$ is given by

$$\hat{S}(x) = \mu + r'V^{-1}(Y - \mu\mathbf{1}) \tag{6.13}$$

where r is a vector with elements $r_i = \rho(||x - x_i||) : i = 1, \ldots, n$. The matrix $V^{-1} = (\tau^2 I + R)^{-1}$ is determined by the model and the data locations x_i, but does not depend on the target location x. It follows that

$$
\begin{aligned}
\hat{S}(x) &= \mu + \sum_{i=1}^{n} a_i(x)(Y_i - \mu) \\
&= \{1 - \sum_{i=1}^{n} a_i(x)\}\mu + \sum_{i=1}^{n} a_i(x)Y_i. \tag{6.14}
\end{aligned}
$$

This shows that the predictor $\hat{S}(x)$ compromises between its unconditional mean μ and the observed data Y, and that the nature of the compromise depends on the target location x, the data locations x_i and the values of the model parameters. We call the $a_i(x)$ the *prediction weights*.

The prediction variance is

$$\text{Var}(S(x)|Y) = 1 - r'(\tau^2 I + R)^{-1}r. \tag{6.15}$$

This depends on the target location x, the data locations x_i and the values of the model parameters, but does not depend on the observed values of the measurements Y_i.

Predictive performance is therefore affected both by the underlying model and by the sampling design. Two basic designs are a regular design, using n locations x_i evenly spaced to form a two-dimensional lattice or, in one dimension, equal subdivisions of an interval, and a random design using n locations independently and uniformly distributed over the study region. For any given values of the model parameters, the density of data-points per unit interval is more important than their absolute number, but for these illustrative examples we have chosen to confound the two by standardising the study region to the unit square or, in one dimension, the unit interval.

6.4.1 The prediction weights

For our first example, we show the predictions weights for three sets of sample locations in the unit square. The results are displayed in Figure 6.1.

Figure 6.1 illustrates a feature of both simple and ordinary kriging, namely that distance from the prediction point is an important, but not the only, ingredient in determining the prediction weight attached to a point in the sampling design. For example, in the two left-hand panels of Figure 6.1 the uppermost of the three points in the sampling design is closest to the target location, and is given the largest weight.

The two central panels of Figure 6.1 illustrate a property of kriging known as *de-clustering*, whereby the kriging algorithm gives reduced weight to individual locations within a spatial cluster. This is a distinctive aspect of the kriging predictor compared with other interpolation methods such as inverse squared distance weighting. The down-weighting of individual points makes intuitive sense in this context because one consequence of the assumed spatial correlation structure of the data is that two closely spaced locations convey little more information than does a single, isolated location; notice in particular that for the sample design shown in the centre panels, the combined weight attached to the pair of closely spaced locations is only slightly greater than the weight attached to the corresponding single location in the left-hand panels.

Finally, the two right-hand panels show the *masking* effect when two sample locations and the target location are collinear, or nearly so; the closer of the two sample locations is given a large, positive weight whilst the more distant, masked location is given a negative weight. In general, masked locations can be given positive, zero or negative weights, depending on the assumed correlation model.

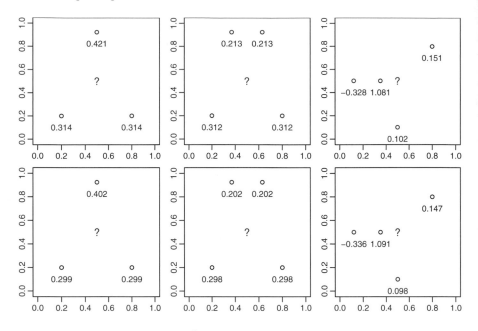

Figure 6.1. Prediction weights for $\hat{S}(0.5, 0.5)$ for three sets of data locations. Gaussian model parameters are $\mu = 0$, $\sigma^2 = 1$, $\tau^2 = 0$, Matérn correlation of order $\kappa = 1.5$ with scale parameter $\phi = 0.1$. Upper panels show the prediction weights using simple kriging with plug-in estimate $\hat{\mu} = \bar{y}$, lower panels show prediction weights using ordinary kriging.

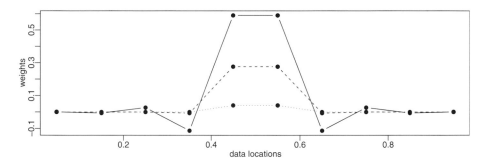

Figure 6.2. Effect of the correlation function parameter ϕ on the kriging weights. Dots indicates the values of the prediction weights for $\hat{S}(0.5)$ using simple kriging with a regular design of 10 sample locations equally spaced on the unit interval. Gaussian model parameters are $\mu = 0$, $\sigma^2 = 1$, $\tau^2 = 0$, Matérn correlation of order $\kappa = 1.5$ with scale parameter $\phi = 0.1$ (solid line), $\phi = 0.02$ (dashed line), $\phi = 0.01$ (dotted line).

For our next example, we consider 10 data locations equally spaced on the unit interval, $x_i = 0.05 + 0.1i : i = 1, \ldots, 10$. Figure 6.2 shows the prediction weights $a_i(x)$ when $x = 0.5$, mid-way between two data locations, using simple kriging with μ treated as known. The model parameters are $\tau^2 = 0$ and Matérn correlation of order $\kappa = 1.5$, with ϕ taking each of the values $\phi = 0.1, 0.02, 0.01$.

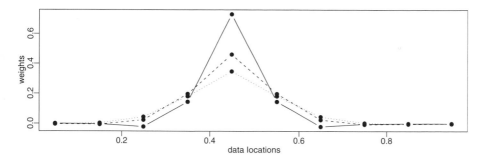

Figure 6.3. Effect of the nugget parameter on the kriging weights. Dots are values of the prediction weights for $\hat{S}(0.45)$ using simple kriging with an equi-spaced design of 10 sample locations on the unit interval. Gaussian model parameters are $\mu = 0$, $\sigma^2 = 1$, $\tau^2 = 0.1$ (solid line), $\tau^2 = 0.5$ (dashed line), $\tau^2 = 1$ (dotted line), Matérn correlation of order $\kappa = 1.5$ with scale parameter $\phi = 0.1$.

In each case, the general pattern is that the largest weights are those associated with data locations x_i immediately either side of the target location x, but the detailed pattern varies with ϕ. Note also that as ϕ decreases, corresponding to generally weaker correlations between $S(x)$ and the Y_i, the sum of the weights decreases. As ϕ approaches zero, the weights also approach zero and $\hat{S}(x) \approx \mu = 0$, because $S(x)$ and Y are then independent, hence the observed values of Y are of no help in predicting $S(x)$.

Figure 6.3 shows the effect of the nugget variance on the pattern of simple kriging weights for predicting $S(x)$ when $x = 0.45$, coinciding with one of the data locations. The model parameters are now $\kappa = 1.5$, $\phi = 0.1$ and $\tau^2 = 0.1, 0.5$ or 1.0. Note firstly that when $\tau^2 = 0$ (not shown), $a_5(x) = 1$, all other $w_i(x) = 0$ and $\hat{S}(x) = Y_5$. This is sensible, because $\tau^2 = 0$ implies that $S(0.45) = Y_5$ exactly. More generally, whenever $\tau^2 = 0$ the simple kriging predictor $\hat{S}(x)$ interpolates the data, i.e. at each sampled location x_i, $\hat{S}(x_i) = Y_i$. As the value of τ^2 increases, the prediction weights are spread progressively over more of the Y_i and the total weight decreases. For very large τ^2, the noise in the data dominates the signal, implying that $S(x)$ and Y are approximately independent, the weights all approach zero and $\hat{S}(x) \approx \mu = 0$, for any x.

Figure 6.4 shows the prediction weights for $\hat{S}(x)$ in a simple two-dimensional example where the target location is surrounded by four data locations. The model is again Matérn with $\kappa = 1.5$ and the three panels of Figure 6.4 correspond to $\phi = 0.2, 0.1$ and 0.05. The prediction algorithm is now ordinary, rather than simple kriging, hence as ϕ approaches zero we obtain $\hat{S}(x) \approx \bar{y}$. Figure 6.5 shows the effect of varying the nugget variance τ^2 in this example. The target for prediction is again the central location $x = (0.5, 0.5)$ and the model Matérn with $\kappa = 1.5$, but now we fix $\phi = 0.1$ and show results for $\tau^2 = 0.1, 0.5$ and 1.0. As τ^2 increases, the ordinary kriging predictor approaches $\hat{S}(x) = \bar{y}$.

Recall that we extend the domain of the correlation function $\rho(u)$ to the real line by defining $\rho(-u) = \rho(u)$. With this extension, the exponential, $\rho(u) = \exp(-u/\phi)$, is continuous but non-differentiable at $u = 0$, whereas the Matérn of order 1.5 is differentiable everywhere. Figure 6.6 shows a set of 10 values Y_i at

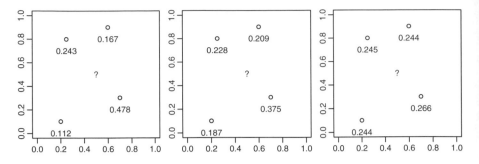

Figure 6.4. Effect of the correlation function parameter ϕ on the kriging weights. Gaussian model parameters are $\mu = 0$, $\sigma^2 = 1$, $\tau^2 = 0$. Panels show prediction weights for $\hat{S}(0.5, 0.5)$ using ordinary kriging with an design of 4 sample locations on the unit square. Matérn correlation of order $\kappa = 1.5$ with scale parameter $\phi = 0.2$ (left panel), $\phi = 0.1$ (middle panel), $\phi = 0.05$ (right panel).

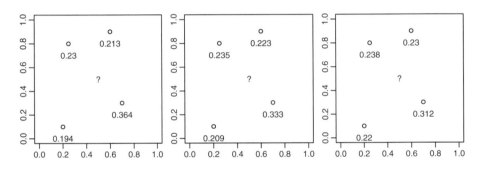

Figure 6.5. Effect of the nugget parameter on the kriging weights. Panels show prediction weights for $\hat{S}(0.5, 0.5)$ using ordinary kriging with an design of 4 sample locations on the unit square. Gaussian model parameters are $\mu = 0$, $\sigma^2 = 1$, $\tau^2 = 0.1$ (left), $\tau^2 = 0.5$ (centre), $\tau^2 = 1$ (right), Matérn correlation of order $\kappa = 1.5$ with scale parameter $\phi = 0.1$.

evenly spaced locations x_i, together with the predictors $\hat{S}(x)$ assuming $\tau^2 = 0$ and either the exponential or the differentiable Matérn correlation function, in each case with $\phi = 0.1$. The predictors inherit the analytic smoothness of the assumed correlation function — continuous for the exponential, differentiable for the Matérn. This suggests that any contextual knowledge concerning the smoothness of the underlying signal should be one consideration in choosing a correlation function for particular applications.

6.4.2   Varying the correlation parameter

Figure 6.7 shows the result of an experiment with nine measurements y_i taken at randomly located points x_i on the unit interval. The Gaussian model parameters are $\mu = 0$, $\sigma^2 = 1$, $\tau^2 = 0$ and a Matérn correlation function with $\kappa = 1.5$ and $\phi = 0.1, 0.025$ or 0.01. As in our previous experiment, in every case the simple kriging predictor $\hat{S}(x)$ interpolates the data, but its behaviour away from the

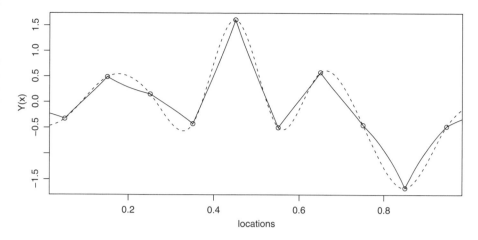

Figure 6.6. The simple kriging predictor for an equi-spaced design of 10 sample locations on the unit interval. Gaussian model parameters are $\mu = 0$, $\sigma^2 = 1$, $\tau^2 = 0$, Matérn correlation of order $\kappa = 0.5$ (solid line) or $\kappa = 1.5$ (dashed line) with scale parameter $\phi = 0.1$.

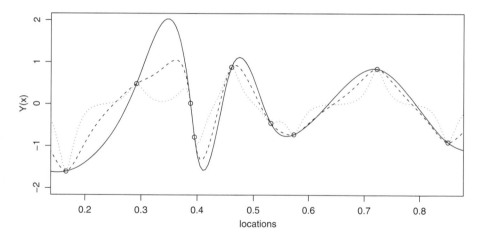

Figure 6.7. The simple kriging predictor for a random design of nine sample locations on the unit interval. Gaussian model parameters are $\mu = 0$, $\sigma^2 = 1$, $\tau^2 = 0$, Matérn correlation of order $\kappa = 1.5$ and scale parameter $\phi = 0.1$ (solid line), $\phi = 0.025$ (dashed line), $\phi = 0.01$ (dotted line).

data locations x_i is affected by the value of ϕ in the following way. Suppose that we wish to predict $S(x)$ at a location remote from *all* of the x_i. Then, for sufficiently small ϕ, the correlation between $S(x)$ and any of the Y_i will be small, and the observed values of Y_i correspondingly of little value in predicting $S(x)$. Thus, $\hat{S}(x)$ will be approximately equal to its unconditional expectation, which in this case is zero. As the prediction location x moves closer to any or all of the data locations, the correlations between $S(x)$ and the Y_i increase,

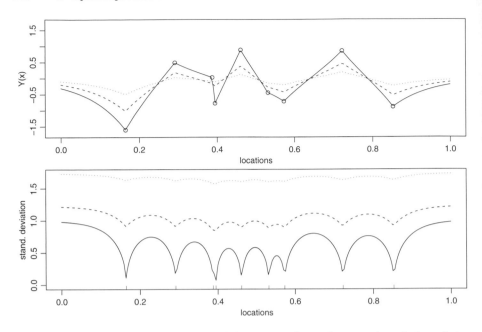

Figure 6.8. Upper panel shows the simple kriging predictor for a random design of nine sample locations on the unit interval. Gaussian model parameters are $\mu = 0$, $\sigma^2 = 1$, $\tau^2 = 0$ (solid line), $\tau^2 = 0.5$ (dashed line), $\tau^2 = 2$ (dotted line), Matérn correlation of order $\kappa = 0.5$ with scale parameter $\phi = 0.1$. Lower panel shows the simple kriging prediction standard errors for the same design of nine sample locations, now indicated by vertical tick-marks above the x-axis.

and the observed values of the Y_i make a correspondingly bigger impact on the predicted surface $\hat{S}(x)$. In general, the predictor $\hat{S}(x)$ is a compromise between the observations Y_i and the prior expectation zero, and the balance between the two depends on the overall correlation structure.

6.4.3 Varying the noise-to-signal ratio

Figure 6.8 involves the same nine data locations and measurements as in the previous example. The upper panel now shows predictions for the Matérn correlation function with $\kappa = 0.5$ (exponential model), $\phi = 0.1$ and each of $\tau^2 = 0, 0.5$ and 2.0. This illustrates that when τ^2 is positive, $\hat{S}(x)$ smooths rather than interpolates, and that larger values of τ^2 give progressively more smoothing towards the unconditional mean, $\mu = 0$. The lower panel of Figure 6.8 shows the prediction standard deviation, $\sqrt{\mathrm{Var}\{S(x)|Y\}}$, as a function of x. The general pattern is that the prediction standard deviation increases with increasing distance from neighbouring data locations, falling to zero at the data locations if and only if $\tau^2 = 0$.

6.5 Trans-Gaussian kriging

The term trans-Gaussian kriging was coined by Cressie (1993) to refer to minimum mean-square error prediction using the transformed Gaussian model described in Section 3.8.

As discussed in Section 3.8, one way to extend the applicability of the Gaussian model is to assume that it holds only when the vector Y is transformed component-wise. Specifically, for a set of data $(x_i, y_i) : i = 1, \ldots, n$, we define $y_i^* = h(y_i)$ for some known function $h(\cdot)$ and assume that the transformed data y_i^* are generated by an underlying Gaussian model. For the time being, we assume that the Gaussian model is stationary, but all of the non-stationary extensions discussed earlier can be applied to transformed data in the obvious way.

Because the transformation function $h(\cdot)$ is assumed known, the results of Section 6.2 apply to prediction of $S(x)$ or other properties of the signal except that Y^* replaces Y throughout. However, when a transformation is used, predictions are usually required on the scale of the original observations, in which case we need to allow for the non-linearity in $h(\cdot)$. The simplest way to formalise this is to assume that the target for prediction is

$$T(x) = h^{-1}\{\mu + S(x)\}. \tag{6.16}$$

In general, evaluation of the minimum mean square error predictor for (6.16) is not straightforward. One exception, and the most common example in practice, is when $h(\cdot) = \log(\cdot)$. Then, $h^{-1}(\cdot) = \exp(\cdot)$ and (6.16) can be written as

$$T(x) = \exp(\mu)\exp\{S(x)\} = \exp(\mu)T_0(x). \tag{6.17}$$

The conditional distribution of $S(x)$ given Y^* is univariate Gaussian, with mean and variance $\hat{S}(x)$ and $v(x)$ given by (6.7) and (6.8) except that Y^* replaces Y. The distribution of $T_0(x) = \exp\{S(x)\}$ under the same conditioning is log-Gaussian, and standard properties of the log-Gaussian distribution imply that

$$\hat{T}_0(x) = \exp\{\hat{S}(x) + v(x)/2\} \tag{6.18}$$

with prediction variance

$$\mathrm{Var}\{\hat{T}_0(x)|Y^*\} = \exp\{2\hat{S}(x) + v(x)\}[\exp\{v(x)\} - 1]. \tag{6.19}$$

Note in particular that the second term within the exponential on the right hand side of (6.18) is non-negative, and can be non-negligible however large the data-set on which it is based.

Other transformation functions $h(\cdot)$ can be handled approximately by using a low-order Taylor series expansion of $h^{-1}(\cdot)$. For example, suppose that the data are expressed as proportions $p_i : i = 1, \ldots, n$. One possible approach to analysing such data is to apply a logit transform to obtain transformed data $y_i^* : i = 1, \ldots, n$, where

$$y_i^* = h(p_i) = \log\{p_i/(1 - p_i)\},$$

with inverse transform

$$p_i = h^{-1}(y_i^*) = \{1 + \exp(-y_i^*)\}^{-1}. \tag{6.20}$$

Analysing the data on the y^*-scale, we might then use a linear Gaussian model to obtain predictions $\hat{S}(x)$ and associated prediction variances $v(x)$ at each location. The target for prediction on the original scale is $T = h^{-1}\{S(x)\}$, for an arbitrary location x. We consider a Taylor series expansion of T about $\hat{S}(x)$. Writing $g(\cdot) = h^{-1}(\cdot)$, and suppressing the dependence on the location x, this gives

$$T \approx g(\hat{S}) + (S - \hat{S})g'(\hat{S}) + 0.5(S - \hat{S})^2 g''(\hat{S}).$$

Now, taking expectations with respect to the conditional distribution of S given the data Y, we obtain the approximation

$$E[T|Y] \approx g(\hat{S}(x)) + 0.5v(x)g''(\hat{S}(x))$$

for the minimum mean square error predictor, $\hat{T} = E[T|Y]$. Substitution of $g(\cdot) = h^{-1}(\cdot)$ from (6.20) gives the explicit expression

$$\hat{T}(x) = \{1 + e(x)\}^{-1} - 0.5v(x)e(x)\{1 - e(x)\}\{1 + e(x)\}^{-3},$$

where $e(x) = \exp\{-\hat{S}(x)\}$. Note that the correction to the naive predicted proportion, $h^{-1}\{\hat{S}(x)\} = \{1 + e(x)\}^{-1}$, is negative if $e(x) < 1$, corresponding to a naive predicted proportion greater than 0.5, and conversely is positive if the naive predicted proportion is less than 0.5.

We now give a simple, one-dimensional illustration of log-Gaussian kriging. The data were generated by a log-Gaussian model for $Y(x)$, assuming an underlying Gaussian process $S(x)$ with $\mu = 0$, $\sigma^2 = 2$, and Matérn correlation function of order $\kappa = 1.5$ with $\phi = 0.15$. The measurements $Y_i^* = \log(Y_i)$ were generated at 11 unequally spaced points on the unit interval, with noise-to-signal variance ratio $\tau^2/\sigma^2 = 0.1$. The solid line in Figure 6.9 shows the realisation of $\exp\{S(x)\}$ along with the data Y indicated by the circles. The figure compares $\exp\{S(x)\}$ with two predictors: the minimum mean square error predictor of $\exp\{S(x)\}$ as defined by (6.18); and the naive predictor $\exp\{\hat{S}(x)\}$ where $\hat{S}(x)$ is the minimum mean square error predictor of $S(x)$. Note that the bias-correction between the naive and minimum mean square error predictors has a noticeable effect in regions where the sampling is sparse, as the correction term involving the prediction variance is then more important.

As discussed in Section 6.2.4, prediction on the untransformed scale formally corresponds to a non-linear target T in the Gaussian model, and closed-form expressions for the minimum mean square error predictor $\hat{T} = E[T|Y]$ can only be found in special cases, as shown above in the case of log-Gaussian kriging. However, it is straightforward (although sometimes computationally expensive) to generate an independent random sample from the predictive distribution of the signal process $S(x)$ on a fine grid to cover the study region. The corresponding values of the target, $T_1, ..., T_s$ say, then form an independent random sample of size s from the predictive distribution of T, and the sample mean $\bar{T}$ gives a Monte Carlo approximation to $\hat{T}$ if required.

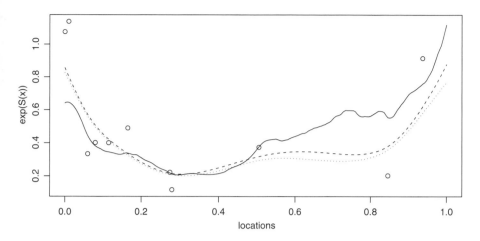

Figure 6.9. A one-dimensional illustration of log-Gaussian kriging. The true signal, $\exp\{S(x)\}$ is shown as a solid line, the observed data Y as a set of open circles. The minimum mean square error predictor is shown as a dashed line and the naive predictor, $\exp\{\hat{S}(x)\}$ as a dotted line.

6.5.1   Analysis of Swiss rainfall data (continued)

In Section 5.4.7 we discussed parameter estimation for the Swiss rainfall data, and concluded that a reasonable model for the data was a Gaussian model on the square-root scale, i.e. a trans-Gaussian model with $h(y) = \sqrt{y}$. For the correlation function we chose the Matérn model with $\kappa = 1$. We now show the resulting spatial prediction of the rainfall surface, using plug-in values of the Gaussian model parameters estimated by maximum likelihood.

Figure 6.10 shows plug-in predictions of $T(x) = S^2(x)$ and the corresponding prediction variances, computed on a regular 7.5×7.5 km grid. Note that prediction variances are large at locations where the predictions themselves are large. This is one consequence of using the transformed model with $\lambda < 1$.

Another potentially interesting target for prediction is the proportion of the total area for which rainfall exceeds some threshold value c. We denote this target by $A(c)$. For illustration, we choose a threshold of $c = 200$. For this non-linear target the plug-in predictor is analytically intractable. We have therefore computed it from independent conditional simulations, generating a Monte Carlo sample of size 1000.

Using the simple Monte Carlo approximation described above, we obtain the plug-in prediction $\tilde{A}(200) = 0.394$. More interestingly, the left-hand panel of Figure 6.11 shows a histogram of a Monte Carlo sample from the predictive distribution of $A(200)$, again based on 1000 simulations. Each simulation contributes a point to the histogram as follows: we generate a realisation from the predictive distribution of $S(x)$ at points x on a fine grid and calculate the approximate value of $A(200)$ for this realisation as the proportion of grid-locations for which the realised $S(x)$ is greater than 200. From the resulting histogram,

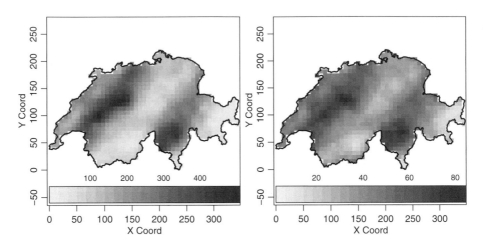

Figure 6.10. Swiss rainfall data. Plug-in predictions of rainfall (left-hand panel) and corresponding prediction variances (right-hand panel).

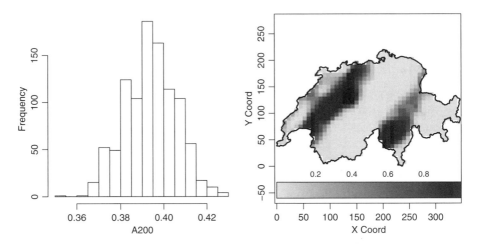

Figure 6.11. Prediction of functionals of $S(x)$ for the Swiss rainfall data. The left-hand panel shows the histogram of a sample of size 1000 drawn from the predictive distribution of $A(200)$. The right-hand panel shows the map of exceedance probabilities $P(x; 250)$.

we can read off predictive probabilities for $A(200)$ to lie within any stated limits, for example $P(0.375 < A(200) < 0.412) = 0.90$.

A further possible prediction target is a map of the probabilities, $P(x; c)$ say, that $S(x)$ exceeds the threshold value c, given the data. The right-hand panel of Figure 6.11 shows a map of $P(x; 250)$, which is obtained by computing for each point in the prediction grid the proportion of simulated values of $S(x)$ which exceed 250.

6.6   Kriging with non-constant mean

Estimation of the spatial trend, $\mu(x) = E[Y(x)]$ where $Y(x)$ represents the response at location x, may be of interest for two different reasons. In some problems, identifying a model for $\mu(x)$ is of direct scientific interest. Typically, this arises when the experimenter has recorded a number of spatial explanatory variables, $d_k(x)$, and wishes to know which of these influence the mean response, $\mu(x)$. In other problems, the scientific goal is the prediction of an underlying surface but a spatial trend is nevertheless evident and by including a term for the trend, typically modelled empirically as a low-degree polynomial trend surface, we improve the precision of the resulting predictions.

Both types of problem are embraced by the Gaussian model with a linear specification for the trend,

$$\mu(x) \quad = \quad \beta_0 + \sum_{k=1}^{p} \beta_k d_k(x) = d(x)'\beta, \qquad (6.21)$$

where $d(x)' = (1, d_1(x), ..., d_p(x))$. Inference about the regression parameters β, either likelihood-based or Bayesian with a pragmatic choice of prior, is relatively straightforward using the methods described in Chapters 5 and 7, respectively. Here, we consider the problem of predicting realised values, or more general properties of an underlying spatial surface.

6.6.1   Analysis of soil calcium data (continued)

Figure 6.12 shows maps of predicted values and the corresponding standard errors for the soil calcium data of Example 1.4. Recall that for these data the study area is divided into three sub-areas according to their management history. As described in Section 5.4.8, our fitted model for these data has mean parameters $(39.71, 47.75, 53.52)$ for the three sub-areas and covariance structure described by a stationary Gaussian process with signal variance $\sigma^2 = 98.7$, exponential correlation function with $\phi = 72.61$ and nugget variance $\tau^2 = 3.26$. The predictions were obtained at points covering the whole of the study area at a spacing of 10 m. Note that the prediction map shows discontinuities at the boundaries between the sub-areas as a consequence of treating sub-area as a three-level factor.

An analysis of this kind requires the explanatory variable to be recorded at both data locations and prediction locations. In classical geostatistical terminology, this is called kriging with an external trend.

6.7   Computation

Prediction as discussed in this chapter is implemented in **geoR** by the function `krige.conv()`. The name is a mnemonic for "conventional kriging," in contrast with another function for geostatistical prediction, `krige.bayes()`, which implements a Bayesian algorithm to be described in Chapter 7. We have already

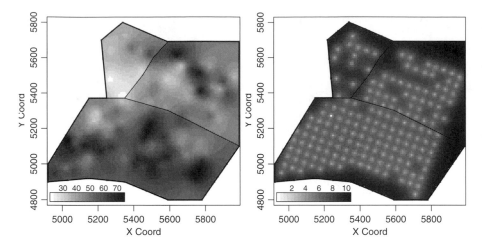

Figure 6.12. Kriging predictions for the soil calcium data, treating sub-area as a factor on three levels. The left-hand panel shows point predictions, the right-hand panel prediction standard errors. See text for model specification.

shown how to use the function `krige.conv()` in Section 2.8, where we applied it to the surface elevation data. Here we examine its options in greater detail, starting from its arguments.

```
> args(krige.conv)
```

```
function (geodata, coords = geodata$coords, data = geodata$data,
    locations, borders, krige, output)
NULL
```

The analysis carried out by a single call to the kriging function `krige.conv()` requires input parameters to define: the data and coordinates using `geodata`; the locations where predictions are required (`locations`); and the model as specified by the argument `krige`. For convenience we pass arguments to `krige` using `krige.control()` to specify the parameters of the model, which can be done either by specifying the individual values for the model parameters, or by an object which stores the results of a model-fitting procedure as presented in Section 5.6. An additional, optional argument `border` allows the definition of the border of the region, which does not affect the calculations, but is particularly useful for displaying results when we are making predictions over non-rectangular regions.

The resulting object contains at least the elements `sk$predict` and `sk$krige.var`, which contain the predicted values and kriging variances for the prediction locations. However, other results can be obtained by optional elements of `output`, which are selected by the call to the `output.control()` function. For example, this function allows us to obtain simulations from the predictive distribution, such as conditional simulations, or estimates of quantiles, percentiles, and so on.

We now show the commands used to obtain the results reported earlier for the Swiss rainfall data. The data are available within **geoR** and are loaded using the data() function. Estimation of model parameters uses the likfit() function as discussed in Chapter 5. In the call below, we set the argument lambda=0.5 to fix the value for the parameter of the Box-Cox transformation.

```
> data(SIC)
> ml <- likfit(sic.all, ini = c(100, 40), nug = 10, lambda = 0.5,
+     kappa = 1)
```

For prediction, we first define a grid of prediction points using pred_grid(). This takes as its main arguments a polygon defining the prediction region, which is typically the border of the study area and the grid spacing. A call to krige.control() then passes the model parameters. In this example we pass the model parameters as a single object, m1, but as discussed above the krige.control() function also allows for the specification of each model parameter individually. By default, krige.control() performs ordinary kriging. For the output, the call to output.control() includes options to generate and store 1000 simulations of the conditional distribution of S given Y and to define a threshold value 250 which will be used to compute exceedance probabilities at each of the prediction locations. We then set the random seed, using set.seed(), so as to allow for reproduction of the simulation results if required. Finally, we call the prediction function krige.conv() according to the description given above.

```
> gr <- pred_grid(sic.borders, by = 7.5)
> KC <- krige.control(obj.model = ml)
> OC <- output.control(n.pred = 1000, simul = TRUE, thres = 250)
> set.seed(2419)
> pred <- krige.conv(sic.all, loc = gr, borders = sic.borders,
+     krige = KC, out = OC)
```

The maps shown in Figure 6.10 are then obtained using the built-in image() function.

```
> image(pred, col = gray(seq(1, 0.1, 1 = 21)), zlim = predlim,
+     x.leg = c(0, 350), y.leg = c(-60, -30))
> image(pred, loc = gr, val = sqrt(pred$krige.var), zlim = selim,
+     col = gray(seq(1, 0.1, 1 = 21)), x.leg = c(0, 350),
+     y.leg = c(-60, -30))
```

Figure 6.11 shows predictions of two functionals of $S(x)$ obtained from the simulated realisations produced by the call to krige.conv(). The first is the predictive distribution of $A(200)$, the proportion of area with rainfall greater than 200. We obtain this by post-processing the simulations using apply() as shown below. The second is the map of exceedance probabilities $P(x; 250)$, which was included in the output by the options set in the call to output.control().

```
> dim(pred$simulations)
> A200 <- apply(pred$simul, 2, function(y) sum(y > 200)/length(y))
> hist(A200, main = "")
```

```
> image(pred, val = 1 - pred$prob, col = gray(seq(0.9, 0.1,
+     l = 41)), x.leg = c(0, 350), y.leg = c(-60, -30))
```

Our second example in this section concerns the soil calcium data. We proceed in a similar manner as for the analysis of the Swiss rainfall data, using the commands given below. We first load the data, which are also included with **geoR**, and fit a model including the covariate **area**.

```
> data(ca20)
> fit <- likfit(ca20, ini = c(100, 60), trend = ~area)
```

For the predictions, we define a grid with spacing of 10 metres between the points using the **pred_grid()** function as in the previous example. Because the area is non-rectangular, we then use the **polygrid()** function to select the grid points which fall within the prediction area. Next, we use this to build a covariate vector with dimension equal to the number of prediction locations indicating to which area each of the prediction points belongs.

```
> gr <- pred_grid(ca20$borders, by = 10)
> gr0 <- polygrid(gr, borders = ca20$border, bound = T)
> ind.reg <- numeric(nrow(gr0))
> ind.reg[.geoR_inout(gr0, ca20$reg1)] <- 1
> ind.reg[.geoR_inout(gr0, ca20$reg2)] <- 2
> ind.reg[.geoR_inout(gr0, ca20$reg3)] <- 3
> ind.reg <- as.factor(ind.reg)
```

We now use the **krige.control()** function and associated methods to implement the predictions and display the results. Since this is a prediction with covariates we pass the covariate values at data locations to the argument **trend.d** and at prediction locations to **trend.l**.

```
> KC <- krige.control(trend.d = ~area, trend.l = ~ind.reg,
+     obj.model = fit)
> ca20pred <- krige.conv(ca20, loc = gr, krige = KC)
> par(mar = c(2.8, 2.5, 0.5, 0.5), mgp = c(1.8, 0.7, 0),
+     mfrow = c(1, 2))
> image(ca20pred, loc = gr, col = gray(seq(1, 0, l = 21)),
+     x.leg = c(4930, 5350), y.leg = c(4790, 4840))
> polygon(ca20$reg1)
> polygon(ca20$reg2)
> polygon(ca20$reg3)
> image(ca20pred, loc = gr, val = sqrt(ca20pred$krige.var),
+     col = gray(seq(1, 0, l = 21)), x.leg = c(4930, 5350),
+     y.leg = c(4790, 4840))
> polygon(ca20$reg1)
> polygon(ca20$reg2)
> polygon(ca20$reg3)
```

Notice that the kriging function **krige.conv()** does not return the kriging weights, although the calculations of these are performed internally. However, the weights can be obtained using the function `krweights()`, which takes as arguments the data coordinates, the location(s) of the point(s) to be predicted and the object which specifies the model. For example, to obtain the weights shown in the lower-left panel of Figure 6.3 we use the commands below.

```
> coords <- cbind(c(0.2, 0.25, 0.6, 0.7), c(0.1, 0.8, 0.9,
+       0.3))
> KC <- krige.control(ty = "ok", cov.model = "mat", kap = 1.5,
+       nug = 0.1, cov.pars = c(1, 0.1))
> krweights(coords, c(0.5, 0.5), KC)

[1] 0.1935404 0.2301559 0.2125838 0.3637199
```

6.8 Exercises

6.1. Evaluate the prediction weights associated with simple kriging, treating all model parameters as known, when the model is a stationary Gaussian process with $\mu = 0$, $\sigma^2 = 0$, $\tau^2 = 0$ and exponential correlation function $\rho(u) = \exp(-u/\phi)$, and the sampling locations are equally spaced along the unit interval. Which of the weights are zero, and why?

6.2. Extend the result of Exercise 6.1 to unequally spaced sampling locations. Do you get the same pattern of zero and non-zero weights? Comment briefly.

6.3. Extend the result of Exercise 6.1 to a two-dimensional set of sampling locations (for ease of calculation, use a single prediction location and a small number of sample locations). Do you get the same pattern of zero and non-zero weights? Comment briefly.

6.4. Consider a stationary trans-Gaussian model with known transformation function $h(\cdot)$, let x be an arbitrary location within the study region and define $T = h^{-1}\{S(x)\}$. Find explicit expressions for $P(T > c|Y)$ where $Y = (Y_1, ..., Y_n)$ denotes the observed measurements on the untransformed scale and:
(a) $h(u) = u$
(b) $h(u) = \log u$
(c) $h(u) = \sqrt{u}$.

6.5. Simulate and display realisations of zero-mean, unit variance stationary Gaussian processes $S(x)$ on a 40 by 40 grid of points in the unit square, experimenting with different values for the correlation parameters to give a range of "rough" and "smooth" surfaces. Note that for all of these processes, the expected proportion of the unit square for which $S(x) > 0$ should be one-half, although the actual proportion will vary between realisations.

For each selected realisation, take as the data a random sample of size n from the 1600 grid-point values of $S(x)$.

(a) Obtain the predictive distribution of the proportion of the study area for which $S(x) > 0$, i.e. $A(0)$ in the notation of Section 6.5.1, using plug-in predictions with:

   (i) true parameter values

   (ii) parameter values estimated by maximum likelihood.

   Compare the two predictive distributions obtained under (i) and (ii).

(b) Investigate how the predictive distributions change as you increase the sample size, n.

(c) Comment generally.

7

Bayesian inference

In Chapters 5 and 6 we discussed geostatistical inference from a classical or non-Bayesian perspective, treating parameter estimation and prediction as separate problems. We did this for two reasons, one philosophical the other practical. Firstly, in the non-Bayesian setting, there is a fundamental distinction between a *parameter* and a *prediction target*. A parameter has a fixed, but unknown value which represents a property of the processes which generate the data, whereas a prediction target is the realised value of a random variable associated with those same processes. Secondly, estimation and prediction are usually operationally separate in geostatistical practice, meaning that we first formulate our model and estimate its parameters, then plug the estimated parameter values into theoretical prediction equations as if they were the true values. An obvious concern with this two-phase approach is that ignoring uncertainty in the parameter estimates may lead to optimistic assessments of predictive accuracy. It is possible to address this concern in various ways without being Bayesian, but in our view the Bayesian approach gives a more elegant solution, and it is the one which we have adopted in our own work.

7.1 The Bayesian paradigm: a unified treatment of estimation and prediction

7.1.1 Prediction using plug-in estimates

In general, a geostatistical model is specified through two sub-models: a sub-model for an unobserved spatial process $\{S(x) : x \in \mathbb{R}^2\}$, called the signal, and a sub-model for the data $Y = (Y_1, \ldots, Y_n)$ conditional on $S(\cdot)$. Using θ as a generic notation for all unknown parameters, a formal notation for the model specification is

$$[Y, S|\theta] = [S|\theta][Y|S, \theta], \tag{7.1}$$

where S denotes the whole of the signal process, $\{S(x) : x \in \mathbb{R}^2\}$. The square bracket notation, $[\cdot]$, means "the distribution of" the random variable or variables enclosed in the brackets, with a vertical bar as usual denoting conditioning. Whilst we find this notation helpful in emphasising the structure of a model, it will sometimes be more convenient to use the notation $p(\cdot)$ to denote probability or probability density, in which case we reserve $\pi(\cdot)$ to denote the Bayesian prior distribution of model parameters.

The classical predictive distribution of S is the conditional distribution $[S|Y, \theta]$, which in principle is obtainable from the model specification by an application of Bayes' Theorem. For any target for prediction, T, which is a deterministic functional of S the predictive distribution for T follows immediately in principle from that of S, although it may or may not be analytically tractable. In either event, to generate a realisation from the predictive distribution $[T|Y, \theta]$ we need only generate a realisation from the predictive distribution $[S|Y, \theta]$ and apply a deterministic calculation to convert from S to T.

A *plug-in* predictive distribution consists simply of treating estimated parameter values as if they were the truth; hence, for any target T the plug-in predictive distribution is $[T|Y, \hat{\theta}]$.

In the special case of the linear Gaussian model as defined in (5.12) and with a prediction target $T = S(x)$ the plug-in predictive distribution is known explicitly. As demonstrated in Section 6.2.1, $[T|Y, \theta]$ is Gaussian with mean

$$\hat{T} = \mathrm{E}[T|Y, \theta] = \mu(x) + r'V(\theta)^{-1}(Y - \mu)$$

and variance

$$\mathrm{Var}[T|Y, \theta] = \sigma^2(1 - r'V(\theta)^{-1}r),$$

where $\mu(x) = d(x)'\beta$ is the n-element vector with elements $\mu(x_i) : i = 1, ..., n$, $\sigma^2 V(\theta) = \mathrm{Var}(Y|\theta)$ as given by (6.6) and r is the n-element vector of correlations with elements $r_i = \mathrm{Corr}\{S(x), Y_i\}$.

These formulae assume that $S(x)$ has zero mean i.e., any non-zero mean is included in the specification of the regression model for $\mu(x)$. When the target depends on both S and the trend, $\mu(x)$, for example when we want to predict $\mu(x) + S(x)$ at an arbitrary location, we simply plug the appropriate point estimate $\hat{\mu}(x)$ into the definition of T. Plug-in prediction often results in optimistic estimates of precision. Bayesian prediction remedies this.

7.1.2   Bayesian prediction

The Bayesian approach to prediction makes no formal distinction between the unobserved signal process S and the model parameters θ. Both are unobserved random variables. Hence, the starting point is a hierarchically specified joint distribution for three random entities: the data, Y; the signal, S; and the model parameters, θ. The specification extends the two-level hierarchical form (7.1) to a three-level hierarchy,

$$[Y, S, \theta] = [\theta][S|\theta][Y|S, \theta], \tag{7.2}$$

where now $[\theta]$ is the prior distribution for θ. In theory, the prior distribution should reflect the scientist's prior opinions about the likely values of θ prior to collection and inspection of the data; in practice, as we discuss below, the prior is often chosen pragmatically.

The Bayesian predictive distribution for S is defined as the conditional distribution $[S|Y]$. This is again obtained from the model specification by an application of Bayes' Theorem, but starting from (7.2) rather than (7.1). This leads to the result

$$[S|Y] = \int [S|Y, \theta][\theta|Y]d\theta, \tag{7.3}$$

showing that the Bayesian predictive distribution is a weighted average of plug-in predictive distributions, in which the weights reflect our posterior uncertainty about the values of the model parameters θ. As with plug-in prediction, the predictive distribution for any target T which is a functional of S follows immediately, as the transformation from S to T is deterministic. In practice, we simulate samples from the predictive distribution of S, and from each such simulated sample we calculate a corresponding sampled value from the predictive distribution of T.

Typically, but not universally, the Bayesian paradigm leads to more conservative predictions in the sense that the resulting predictive distribution $[T|Y]$ is more dispersed than the plug-in predictive distribution $[T|Y, \hat{\theta}]$. Note also that as the data become more abundant, then for any parameter θ which is identifiable from the data we expect the posterior distribution $[\theta|Y]$ to become progressively more concentrated around a single value $\hat{\theta}$. In other words, the Bayesian predictive distribution for S, and therefore for any target T, converges to the plug-in. However, the rate of convergence is problem specific, depending on a complex inter-play involving the prior, the model and the sampling design. In our experience the difference between the two can be substantial, especially for non-linear targets T. Also, we re-emphasise our point of view that the complete solution to a predictive problem is a probability distribution, not a single value. In geostatistical applications where prediction is the scientific goal, point estimates of parameters may be acceptable, but point predictions are of limited value.

In the special case of the linear Gaussian model with target $T = S(x)$ and pragmatic prior assumptions, we can obtain explicit results for the Bayesian predictive distribution of T. As in Section 5.3, we first illustrate the general approach for the unrealistic case in which all model parameters other than the mean and variance are assumed known, then relax these assumptions to derive a prediction algorithm for the case of practical interest, in which all parameters are assumed unknown and are assigned a joint prior distribution.

7.1.3   Obstacles to practical Bayesian prediction

There are two major requirements which must be met before Bayesian inference can be used in practice.

The first, and the more fundamental, is that the data analyst must be prepared to specify a prior distribution for θ. Often, this is done pragmatically. A guiding principle in applied work is that in the absence of clearly articulated prior knowledge priors should be diffuse, in which case their effects are swamped by the effect of the likelihood provided that the sample size is sufficiently large. However, in the geostatistical setting where data are often highly correlated, intuitive ideas of what constitutes a "large" sample may be misleading. Our experience has been that with data-sets of size several hundred, apparently diffuse priors can still have a noticeable influence on the inferences. It seems to be a general feature of geostatistical problems that the models are poorly identified, in the sense that widely different combinations of parameter values lead to very similar fits. This may not matter if parameter estimation is not of direct interest but is only a means towards the goal of prediction. Even so, it remains a lingering concern because the prior does potentially influence the predictive distribution which we report for any target.

Another issue with regard to prior specification is whether priors for different parameters should be independent. In practice, independent priors are often assumed. However, this assumption is equally often questionable. Note in particular that the substantive meaning of an independent prior specification changes if the model is re-parameterised by anything other than a component-wise transformation.

The second, computational requirement is evaluation of the integral which is required to convert a specified model and prior into a posterior or Bayesian predictive distribution; see (7.4) below. In particular cases, including the linear Gaussian model, it is possible to choose a convenient prior, called the conjugate prior, so that the required integration can be performed analytically. More often, numerical evaluation is required. In most practical problems the only feasible evaluation strategies involve Monte Carlo methods, including the now ubiquitous Markov chain Monte Carlo (MCMC) methods as discussed in Gilks et al. (1996). Conversely, advances in computing power and theoretical developments in Monte Carlo methods of inference have together made Bayesian inference a feasible, and sometimes the only feasible, approach to inference for problems involving complex stochastic models.

7.2   Bayesian estimation and prediction for the Gaussian linear model

We first describe an implementation of Bayesian inference for parameter estimation in the Gaussian linear model. We have argued that parameter estimation is often not the primary goal of a geostatistical analysis. We discuss parameter estimation here as a prelude to the main focus of this chapter, namely Bayesian prediction.

In Bayesian parameter estimation, the likelihood function $\ell(\theta; y)$ is combined with a prior distribution $\pi(\theta)$ via Bayes' Theorem to yield a posterior distribution for θ with density

$$p(\theta|y) = \frac{\ell(\theta; y)\pi(\theta)}{\int \ell(\theta; y)\pi(\theta)d\theta}. \tag{7.4}$$

Inferences about θ are then expressed as probability statements derived from the posterior. For example, the classical notion of a confidence interval for a single parameter, θ_k say, is replaced by a Bayesian credible interval (a, b), where a and b are chosen so that under the posterior distribution for θ, $P(a \leq \theta_k \leq b)$ attains a specified value, for example 0.95. As is the case for a confidence interval, the choice of a and b to achieve a given coverage probability is not unique. Unless stated otherwise, we use a "central quantile-based" method so that, for example, for a 95% credible interval we take a and b to be the 2.5% and 97.5% quantiles of the posterior or predictive distribution. If a point estimate is required, an appropriate summary statistic can be calculated from the posterior, for example its mean, median or mode.

7.2.1 Estimation

We again consider the Gaussian model (5.12) which includes a linear regression specification for the spatial trend, so that

$$[Y] \sim N(D\beta, \sigma^2 R(\phi) + \tau^2 I).$$

To this model specification, whenever possible we add pragmatic specifications for the prior distributions which allow us to obtain explicit expressions for the corresponding posteriors. In other cases we discretise the prior to ease the resulting computations.

We first consider the situation in which we fix $\tau^2 = 0$ i. e., we assume that there is no nugget effect, and all other parameters in the correlation function have known values. Using particular prior specifications, we can then derive the posterior distributions for β and σ^2 analytically. These assumptions are of course unrealistic. We use them simply as a device to lead us towards a feasible implementation in the more realistic setting when all parameters are unknown.

For fixed ϕ, the conjugate prior family for (β, σ^2) is the Gaussian-Scaled-Inverse-χ^2. This specifies priors for β and σ^2 with respective distributions

$$[\beta|\sigma^2, \phi] \sim N\left(m_b, \sigma^2 V_b\right) \quad \text{and} \quad [\sigma^2|\phi] \sim \chi^2_{ScI}\left(n_\sigma, S_\sigma^2\right),$$

where a $\chi^2_{ScI}(n_\sigma, S_\sigma^2)$ distribution has probability density function

$$\pi(z) \propto z^{-(n_\sigma/2+1)} \exp(-n_\sigma S_\sigma^2/(2z)), \quad z > 0. \tag{7.5}$$

As a convenient shorthand, we write this as

$$[\beta, \sigma^2|\phi] \sim N\chi^2_{ScI}\left(m_b, V_b, n_\sigma, S_\sigma^2\right). \tag{7.6}$$

Note, incidentally, that this is one case where a particular form of dependent prior specification is convenient.

Using Bayes' Theorem, we combine the prior with the likelihood given by (5.13) and obtain the posterior distribution of the parameters as

$$[\beta, \sigma^2 | y, \phi] \sim N\chi^2_{ScI}\left(\tilde{\beta}, V_{\tilde{\beta}}, n_\sigma + n, S^2\right), \tag{7.7}$$

where $\tilde{\beta} = V_{\tilde{\beta}}(V_b^{-1}m_b + D'R^{-1}y)$, $V_{\tilde{\beta}} = (V_b^{-1} + D'R^{-1}D)^{-1}$ and

$$S^2 = \frac{n_\sigma S_\sigma^2 + m_b' V_b^{-1} m_b + y'R^{-1}y - \tilde{\beta}' V_{\tilde{\beta}}^{-1} \tilde{\beta}}{n_\sigma + n}. \tag{7.8}$$

Under the conjugate specification, the degree to which the priors influence the inferences for β and σ^2 is controlled by the values of the constants m_b, V_b, n_σ and S_σ^2. Note in particular that the prior mean for $1/\sigma^2$ is $1/S_\sigma^2$ and that the prior distribution for σ^2 becomes less diffuse as n_σ increases. In practice, it may be difficult to elicit appropriate values for these quantities, but in qualitative terms we can think of S_σ^2 as a prior guess at the value of σ^2, and n_σ as a measure of how well informed we consider this prior guess to be. Similarly, the prior mean for β is m_b and its prior distribution becomes less diffuse as V_b decreases, hence the values of m_b and of the elements of V_b should, roughly speaking, reflect our prior guesses and the confidence we wish to place in them.

An alternative prior, often used as a default in Bayesian analysis of linear models, is $\pi(\beta, \sigma^2) \propto 1/\sigma^2$; see for example, O'Hagan (1994). This is an improper distribution, because its integral over the parameter space is infinite. Nevertheless, formal substitution of $V_b^{-1} = 0$ and $n_\sigma = 0$ into the formula (7.7) for the posterior distribution gives the correct expression for the posterior distribution corresponding to this default prior, except that the degrees of freedom are $n - p$, where p is the dimension of β, rather than $n + n_\sigma$.

More realistically, we now allow for uncertainty in all of the model parameters, still considering the case of a model without a nugget effect, so that $\tau^2 = 0$, and with a single correlation parameter ϕ. We adopt a prior $[\beta, \sigma^2, \phi] = [\beta, \sigma^2 | \phi] [\phi]$, the product of (7.6) and a proper density for ϕ. In principle, the prior distribution for ϕ should have continuous support, but in practice we always use a discrete prior, obtained by discretising the distribution of ϕ in equal width intervals. This requires us in particular to specify the range of the prior for ϕ. In the absence of informed scientific opinion, we would do this conservatively, but check that the posterior for ϕ assigns negligible probabilities to the extreme points in the specified prior range.

The posterior distribution for the parameters is then given by

$$[\beta, \sigma^2, \phi | y] = [\beta, \sigma^2 | y, \phi] [\phi | y]$$

with $[\beta, \sigma^2 | y, \phi]$ given by (7.7) and

$$p(\phi | y) \propto \pi(\phi) |V_{\tilde{\beta}}|^{\frac{1}{2}} |R|^{-\frac{1}{2}} (S^2)^{-(n+n_\sigma)/2}, \tag{7.9}$$

where $V_{\tilde{\beta}}$ and S^2 are given by (7.7) and (7.8) respectively. When the prior is $\pi(\beta, \sigma^2, \phi) \propto \pi(\phi)/\sigma^2$, the equation above holds with $n_\sigma = -p$.

Berger, De Oliveira and Sansó (2001) use a special case of this as a default non-informative prior for the parameters of a spatial Gaussian process.

To simulate samples from this posterior, we proceed as follows. We apply (7.9) to compute posterior probabilities $p(\phi|y)$, noting that in practice the support set will be discrete. We then simulate a value of ϕ from $[\phi|y]$, attach the sampled value to $[\beta, \sigma^2|y, \phi]$ and obtain a simulation from this distribution. By repeating the simulation as many times as required, we obtain a sample of triplets (β, σ^2, ϕ) from the joint posterior distribution of the model parameters.

To accommodate a positive nugget variance, $\tau^2 > 0$, in practice we use a discrete joint prior for ϕ and ν^2, where $\nu^2 = \tau^2/\sigma^2$. This adds to the computational load, but introduces no new principles. In this case we replace R in the equations above by $V = R + \nu^2 I$. Similarly, if we wish to incorporate additional parameters into the covariance structure of the signal process $S(\cdot)$, we would again use a discretisation method to render the computations feasible.

Note that the form of Monte Carlo inference used here is direct simulation, replicated independently, rather than MCMC. Hence, issues of convergence do not arise and the simulation-induced variance in sampling from the posterior for any quantity of interest is inversely proportional to the number of simulated replicates. This allows us to assess the magnitude of the simulation-induced variation in the estimated posterior, or summaries of it, and to adjust the number of simulations if necessary.

7.2.2 Prediction when correlation parameters are known

To extend the above results on Bayesian estimation to spatial prediction under the linear Gaussian model, we temporarily assume that all parameters in the correlation function have known values. In other words, we allow for uncertainty only in the parameters β and σ^2.

For fixed ϕ, the conjugate prior family for (β, σ^2) is the Gaussian-Scaled-Inverse-χ^2, and the resulting posterior distribution of the parameters β and σ^2 is given by equations (7.7) and (7.8). The additional step required for prediction is to compute the Bayesian predictive distribution of the signal at an arbitrary set of locations, say $S^* = (S(x_{n+1}), \ldots, S(x_{n+q}))$. This requires us to evaluate the integral,

$$p(s^*|y) = \int \int p(s^*|y, \beta, \sigma^2)\, p(\beta, \sigma^2|y)\, d\beta d\sigma^2, \tag{7.10}$$

where $p(S^*|Y, \beta, \sigma^2)$ is a multivariate Gaussian density with mean and variance given by the extension of (6.7) and (6.8) to the case of a linear regression model for $\mu(x)$. Hence, the mean becomes

$$E[S^*|Y, \beta, \sigma^2] = D^*\beta + r'V^{-1}(Y - D\beta) \tag{7.11}$$

where $V = R + \nu^2 I$ whilst D^* and D are the matrices of covariates corresponding to prediction locations and sampling locations, respectively. The prediction variance is unchanged,

$$\mathrm{Var}[S^*|Y, \beta, \sigma^2] = \sigma^2(1 - r'V^{-1}r). \tag{7.12}$$

The integration in (7.10) yields a q-dimensional multivariate-t distribution defined by:

$$
\begin{aligned}
[S^*|y] &\sim t_{n_\sigma+n}\left(\mu^*, S^2\Sigma^*\right), & (7.13)\\
\mathrm{E}[S^*|y] &= \mu^*,\\
\mathrm{Var}[S^*|y] &= \frac{n_\sigma+n}{n_\sigma+n-2}\,S^2\Sigma^*,
\end{aligned}
$$

where S^2 is given by (7.8) and μ^* and Σ^* by the formulae

$$
\begin{aligned}
\mu^* &= (D^* - r'V^{-1}D)V_{\tilde\beta}V_b^{-1}m_b\\
&\quad + \left[r'V^{-1} + (D^* - r'V^{-1}D)V_{\tilde\beta}D'V^{-1}\right]y,\\
\Sigma^* &= V^0 - r'V^{-1}r + (D^* - r'V^{-1}D)(V_b^{-1} + V_{\hat\beta}^{-1})^{-1}(D^* - r'V^{-1}D)'.
\end{aligned}
$$

The three components in the formula for the prediction variance Σ^* can be interpreted as the variability *a priori*, the reduction due to the conditioning on the data and the increase due to uncertainty in the value of β, respectively.

When it is difficult to elicit informative priors, we would usually adopt diffuse, and apparently uninformative priors. However, the cautionary remarks given in Section 7.1.3 still apply. Prediction of non-linear targets is carried out using a simulation-based sampling procedure similar to the one described in Section 6.2.4.

7.2.3   Uncertainty in the correlation parameters

More realistically, we now allow for uncertainty in all of the model parameters. As in Section 7.2.1, it is helpful first to consider the case of a model without measurement error i.e., $\tau^2 = 0$, and a single correlation parameter ϕ.

We adopt a prior $\pi(\beta, \sigma^2, \phi) = \pi(\beta, \sigma^2|\phi)\,\pi(\phi)$, the product of (7.6) and an independent prior distribution for ϕ, which in practice we specify as a discrete distribution spanning what is thought to be a reasonable range.

The posterior distribution for the parameters is then

$$
[\beta, \sigma^2, \phi|y] = [\beta, \sigma^2|y, \phi]\,[\phi|y]
$$

where $[\beta, \sigma^2|y, \phi]$ is given by (7.7), whilst the posterior density for ϕ is

$$
p(\phi|y) \propto \pi(\phi)\,|V_{\tilde\beta}|^{\frac{1}{2}}\,|R|^{-\frac{1}{2}}\,(S^2)^{-(n+n_\sigma)/2}, \qquad (7.14)
$$

with $V_{\tilde\beta}$ and S^2 given by (7.7) and (7.8), respectively. To simulate samples from this posterior, we again use the simulation method described in Section 6.2.4.

The predictive distribution for the value, $S^* = S(x^*)$ say, of the signal process at an arbitrary location x^*, is given by

$$
\begin{aligned}
[S^*|y] &= \int\!\!\int\!\!\int [S^*, \beta, \sigma^2, \phi|y] \, d\beta \, d\sigma^2 \, d\phi \\
&= \int\!\!\int\!\!\int [S^*, \beta, \sigma^2|y, \phi] \, d\beta \, d\sigma^2 \, [\phi|y] \, d\phi \\
&= \int [S^*|y, \phi] \, [\phi|y] \, d\phi.
\end{aligned}
$$

The discrete prior for ϕ allows analytic calculation of the moments of this predictive distribution. For each value of ϕ we compute the moments of the multivariate-t distribution (7.13) and calculate their weighted sum with weights given by the probabilities $p(\phi|y)$.

To sample from the predictive distribution of S^*, we proceed as follows. We first compute the posterior probabilities $p(\phi|y)$ on the discrete support set of the prior for ϕ, then simulate values of ϕ from the posterior, $[\phi|y]$. For each sampled value of ϕ, we then simulate a value of (β, σ^2) from the Gaussian conditional distribution $[\beta, \sigma^2|y, \phi]$, followed by a value of S^* from the conditional distribution $[S^*|\beta, \sigma^2, \phi, y]$. The resulting value of S^* is an observation from the required predictive distribution, $[S^*|y]$. The same method applies in principle to the simulation of a vector of values in S^* representing $S(x)$ at a number of different prediction points using the multivariate Gaussian predictive distribution $[S^*|y, \beta, \sigma^2, \phi]$. In practice, this may be computationally demanding if the dimensionality of S^* is large.

Finally, when $\tau^2 > 0$ or if we need to introduce additional parameters into the covariance structure of S, we again proceed as in Section 7.2.1 by specifying a discrete joint prior for $[\phi, \nu^2]$, where $\nu^2 = \tau^2/\sigma^2$ and ϕ may now be vector valued.

7.2.4   Prediction of targets which depend on both the signal and the spatial trend

When the target for prediction depends on both the signal process, $S(\cdot)$, and the trend, $\mu(\cdot)$, we need to make a straightforward modification to the Bayesian prediction methodology. At this point, it is helpful to consider the generic notation for the model as given by (7.1), namely $[Y, S|\theta] = [S|\theta][Y|S, \theta]$, but to partition the parameter vector θ as $\theta = (\alpha, \beta)$, where α parameterises the covariance structure and β parameterises the trend. Under this parameterisation, S is conditionally independent of β given α. To sample from the predictive distribution of any target T which depends on both $S(\cdot)$ and μ, we proceed as described in Section 7.2.3, simulating values of α from the distribution $[\alpha|y]$ and of S from the distribution $[S|\alpha, y]$. At this point, if the target T involves the values of the signal, S^* say, at unsampled locations x, we can simulate these directly from the multivariate Gaussian distribution $[S^*|S, \alpha]$, which does not depend on y. Using the same values of α we then simulate values of β, and hence of $\mu(x)$ for any locations x of interest, from the posterior distribution $[\beta|\alpha, y]$. Finally, we use the sampled values of S, S^* and $\mu(x)$ to calculate directly the corresponding sampled values of T which are then realisations from the appropriate predictive distribution, $[T|y]$, as required.

We again emphasise that the simulated samples generated by the methods described in this chapter are exact, independent samples from the required predictive distributions, and the size of the simulation-induced variability can therefore be assessed directly. As we shall discuss in Section 7.5, when we move beyond the linear Gaussian setting, we need to resort to Markov chain Monte Carlo methods, and to address issues concerning the convergence of sampled values to their equilibrium distributions.

7.3   Trans-Gaussian models

A possible approach to the transformed Gaussian model is to consider a parametric family of transformations, such as the Box-Cox family, and to treat the choice of transformation as an additional parameter, λ say, to be estimated.

De Oliveira, Kedem and Short (1997) proceed in this way, using formal Bayesian inference on all of the model parameters. We would be reluctant to follow their approach, partly for the reasons given in Section 6.5, but more particularly because we have reservations about combining predictions using different measurement scales.

If empirical prediction is the sole aim, a possible strategy is the following. Consider a small number of candidate values for λ, for example $\lambda = 1$, 0.5 or 0, and adopt a discrete prior over these candidate values. Now, choose a sensible prior for the remaining parameters conditional on each candidate value of λ, taking care to make these conditional priors at least qualitatively consistent with each other; for example, if the prior for the mean response parameter μ given $\lambda = 1$ is centered on a value around 100, say, then the prior for μ given $\lambda = 0.5$ should be centered around a value of about 10. Predictions can then be made by applying a simple form of Bayesian model averaging, in which the predictive distribution is a mixture of predictive distributions conditional on each candidate value of λ, and weighted according to the posterior probabilities determined for each candidate value.

When the scientific objectives extend beyond empirical prediction, we would prefer to choose λ informally, then carry out a Bayesian analysis treating λ as a fixed, pre-specified quantity, so as to preserve a physical interpretation for each of the model parameters.

There are also some technical objections to applying formal Bayesian inference to the transformed Gaussian model using the Box-Cox family of transformations. Firstly, when using the Box-Cox transformed Gaussian model with $\lambda > 0$, we can back-transform predictions to the original scale using formulae for the moments of the t-distribution. However, this breaks down when $\lambda = 0$, corresponding to the widely used log-transformation, because the exponential of a t-distribution does not have finite moments, hence when $\lambda = 0$ the conditional expectation which usually defines the minimum mean square error predictor does not exist. A second concern is that if $Y > 0$, as is strictly necessary for the Box-Cox transformation to define a real-valued Y^* for all real λ, then Y^* cannot strictly be Gaussian. Of course, this last point applies equally to many situations in which the Gaussian is used routinely as an approximate model for strictly non-negative data.

These comments are intended only as a gentle caution against the unthinking application of the transformed Gaussian model. In practice, the transformed Gaussian is a very useful extension of the Gaussian model when the data are clearly non-Gaussian, and neither the sampling mechanism underlying the data nor the particular scientific context in which the data arise suggests a specific non-Gaussian alternative. Working within the transformed Gaussian framework is also relatively straightforward computationally, by comparison with the perhaps more elegant setting of generalized linear geostatistical models for non-Gaussian data.

7.4 Case studies

7.4.1 Surface elevations

We first consider the surface elevation data from Example 1.1. In Chapter 5 we fitted a linear Gaussian model to these data, assuming a linear trend surface on the coordinates for the mean and a Matérn correlation structure with $\kappa = 1.5$ for the stationary process $S(\cdot)$. We now use this model for spatial prediction of the elevation surface, $T(x) = \mu(x) + S(x)$, and compare the results obtained by plug-in and Bayesian methods.

For plug-in prediction, maximum likelihood estimates of the remaining parameters in the covariance structure are $\hat{\tau}^2 = 34.9$, $\hat{\sigma}^2 = 1693.1$ and $\hat{\phi} = 0.8$. The parameter estimates which define the linear trend surface are $(912.5, -5, -16.5)$. The top-left panel of Figure 7.1 shows the resulting plug-in prediction of surface elevations. The top-right and bottom-right panels show the decomposition of the predicted surface into its two components, the linear trend, $\hat{\mu}(\cdot)$ and the stochastic component $\hat{S}(\cdot)$. The bottom-left panel shows prediction standard errors.

For Bayesian prediction under the same model, we assign the priors as discussed in Section 7.2.1. Specifically, in this example we set $\pi(\beta, \sigma^2) = 1/\sigma^2$ and a joint discrete prior $\pi(\phi, \nu^2) = 1/\phi$ in the region $[0.2, 6] \times [0, 1]$ with 30×21 support points. Figure 7.2 shows prior and posterior distributions for the model parameters ϕ and ν^2. In both cases, the posterior assigns higher probabilities to the lower values of the corresponding parameter.

For comparison with the plug-in results shown in Figure 7.1, we compute the posterior means for $\mu(x)$ and $T(x) = S(x)$ at each location x in a fine grid, and the posterior standard deviations of $T(x)$. The results are shown in Figure 7.3 where, to allow direct comparison, we have used the same grey scales for plug-in and Bayesian maps of prediction results.

Figure 7.4 compares the plug-in and Bayesian predictive distributions at two points. The first, $(x, y) = (5.4, 0.4)$, coincides with one of the data locations

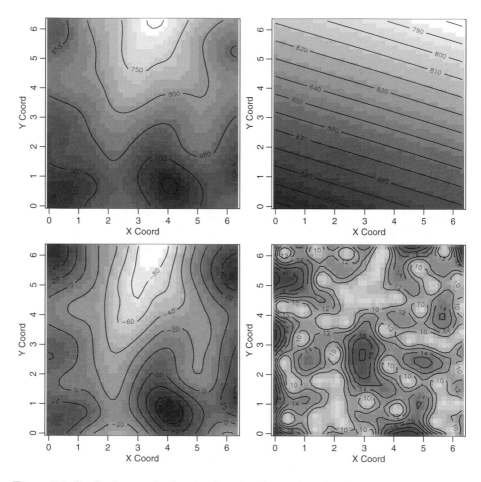

Figure 7.1. Prediction results for the elevation data using plug-in parameter estimates. Top-left panel shows predicted values for $T(x) = \mu(x) + S(x)$, top-right and bottom-left panels shows the two components separately. Bottom-right panel shows the prediction standard errors.

whereas the second, $(x, y) = (1.7, 0.7)$, lies a distance 0.32 from the closest data location. The Bayesian predictive distribution on the left reduces almost to a spike because of the high posterior density at $\nu^2 = 0$, for which the variance of the predictive distribution is equal to zero. Hence, sampled values from the predictive distribution coincide with the observed elevation at this location. For the plug-in prediction, because $\hat{\nu}^2 > 0$ the plug-in predictive distribution is more diffuse. For the other location, the Bayesian predictive distribution has a slightly larger variance than the corresponding plug-in predictive distribution, reflecting the effect of parameter uncertainty. Bayesian predictive distributions typically have larger variances than the corresponding plug-in distributions, but as the example shows, this is not invariably the case.

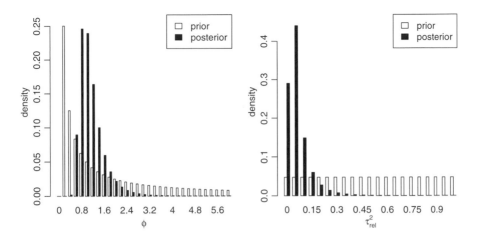

Figure 7.2. Prior and posterior distributions for the model parameters ϕ on the left panel and ν^2 on the right panel.

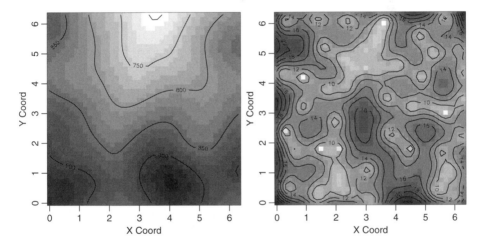

Figure 7.3. Bayesian prediction results for the elevation data. Left panel shows predicted values for $T(x) = \mu(x) + S(x)$ and right panel shows the prediction standard errors.

7.4.2 Analysis of Swiss rainfall data (continued)

In Chapter 5 we obtained the maximum likelihood parameter estimates for the Swiss rainfall data assuming a transformed Gaussian model with transformation parameter $\lambda = 0.5$, constant mean and Matérn correlation function with $\kappa = 1$. In Section 6.5.1 we used the fitted model to obtain plug-in predictions over Switzerland as shown in Figure 6.10.

We now revise the analysis by adopting the Bayesian approach, assuming the same model as before but setting prior distributions for the model parameters $(\mu, \sigma^2, \phi, \nu^2)$ as discussed in Section 7.2.3. For the correlation function parameter ϕ we adopt a reciprocal prior $\pi(\phi) = 1/\phi$ with ϕ taking values on a discrete

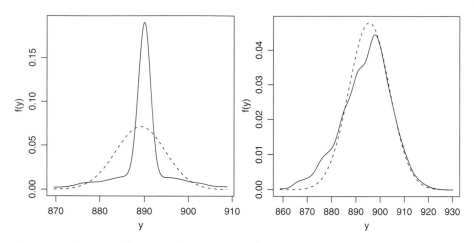

Figure 7.4. Bayesian (solid line) and plug-in (dashed line) predictive distributions at a data location $(x, y) = (5.4, 0.4)$ and a prediction location $(x, y) = (1.7, 0.7)$.

Table 7.1. Swiss rainfall data: posterior means and 95% central quantile-based credible intervals for the model parameters.

Parameter	Estimate	95% interval
β	144.35	[53.08 , 224.28]
σ^2	13662.15	[8713.18 , 27116.35]
ϕ	49.97	[30 , 82.5]
ν^2	0.03	[0 , 0.05]

support of 20 points in the interval $[7.5, 150]$. For the noise to signal variance ratio $\nu^2 = \tau^2/\sigma^2$ we use a uniform prior on a discrete support of 11 points in the interval $[0, 0.5]$. We obtain 1000 samples from posterior and predictive distributions.

Figure 7.5 shows the discrete prior and posterior distributions for the parameters ϕ and ν^2. Table 7.1 shows the 95% credibility intervals for the model parameters. To obtain predictions over the whole of Switzerland we define a 7.5×7.5 km grid of locations. Figure 7.6 shows the predicted values, which range from 2.1 to 499.7, and associated standard errors ranging from 19.3 to 82.9. The limits for the grey scale are the same as were used in the corresponding plot in Figure 6.10, where predicted values ranged from 3.5 to 480.4 and predicted standard errors ranged from 4.8 to 77.5.

We also obtain Bayesian prediction of the proportion of the total area $A(200)$ for which rainfall exceeds the threshold 200 and compare the result with that obtained in Section 6.5.1. We obtain a posterior mean $\tilde{A}(200) = 0.409$. From the sample, we can read off predictive probabilities for $A(200)$ to lie within any stated limits, for example $P(0.391 < A(200) < 0.426) = 0.90$. Recall that for the plug-in predictions the corresponding results were $\tilde{A}(200) = 0.394$ and $P(0.375 < A(200) < 0.41) = 0.90$. The solid line in the left-hand panel of Figure 7.7 shows a density estimate obtained using samples from the predictive

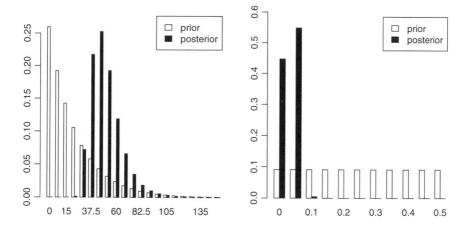

Figure 7.5. Swiss rainfall data: discrete prior and posterior distributions for the parameters ϕ on the left panel and ν^2 on the right panel.

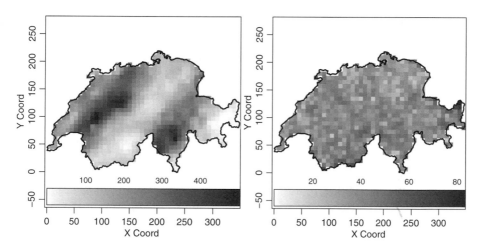

Figure 7.6. Swiss rainfall data. Bayesian predictions of rainfall (left-hand panel) and corresponding prediction variances (right-hand panel).

distribution of $A(200)$ whilst the dashed line shows the corresponding result using plug-in prediction. This illustrates that Bayesian and plug-in methods can give materially different predictive distributions. In our experience, this is especially so for non-linear prediction targets.

Another possible prediction target is a map of the probabilities, $P(x; c)$ say, that $S(x)$ exceeds the threshold value c, given the data. The right-hand panel of Figure 7.7 shows a map of $P(x; 250)$, which is obtained by computing for each point in the prediction grid the proportion of simulated values of $S(x)$ which exceed the value of 250. This result is the Bayesian counterpart of the one obtained for plug-in prediction, as shown in Figure 6.11.

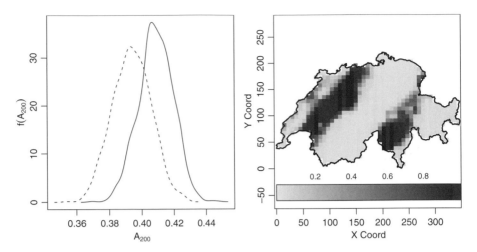

Figure 7.7. Prediction of functionals of $S(x)$ for the Swiss rainfall data. The left–hand panel shows density estimates based on a sample of size $s = 2000$ drawn from the predictive distribution of $A(200)$ using either Bayesian prediction (solid line) or plug-in prediction (dashed line). The right-hand panel shows the map of exceedance probabilities $P(x; 250)$ for the Bayesian prediction.

7.5   Bayesian estimation and prediction for generalized linear geostatistical models

As previously discussed in Section 5.5, the implementation of a likelihood-based method of inference for generalized linear geostatistical models is hampered by the need to evaluate intractable, high-dimensional integrals. For Bayesian inference, the usual way round this difficulty is to use Monte Carlo methods, in particular Markov chain Monte Carlo, to generate samples from the required posterior or predictive distributions. We therefore begin this section with a brief discussion of Markov chain Monte Carlo methods as they apply in the current context. Readers who wish to study Markov chain Monte Carlo methods in detail may want to consult the textbooks listed at the end of Section 1.5.2.

7.5.1   Markov chain Monte Carlo

Markov chain Monte Carlo (MCMC) is now very widely used in Bayesian inference. Its attraction is that, in principle, it provides a way of circumventing the analytical and numerical intractability of Bayesian calculations by generating samples from the posterior distributions associated with almost arbitrarily complex models. MCMC achieves this by simulating from a Markov chain constructed in such a way that the equilibrium distribution of the chain is the required posterior, or Bayesian predictive distribution. Furthermore, it is possible to define general constructions for chains which meet this basic requirement. However, for many applications, constructing reliable MCMC algorithms is difficult. By "reliable" we mean that the chain reaches at least a close approximation to its equilibrium distribution sufficiently quickly to be within the capacity of

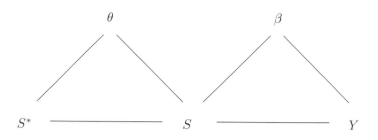

Figure 7.8. Conditional independence graph for the generalized linear geostatistical model. If two nodes are not connected by a line then the corresponding random variables are conditionally independent given the random variables at all other nodes.

the available computing resource and, crucially, that we can recognise when it has reached a close-to-equilibrium state.

We denote by θ the set of parameters which define the covariance structure of the model, and by β the regression parameters which, with $S(\cdot)$, determine the conditional expectation of Y. Our parameterisation assumes that $E[S(x)] = 0$, hence β always includes an intercept term. We write S for the vector of values of $S(x_i)$ at data locations x_i, Y for the corresponding vector of measurements Y_i, and S^* for the vector of values of $S(x)$ at prediction locations x. Note that in practice, the prediction locations may or may not include the data locations x_i. We shall assume that S^* and S are distinct. However, the algorithms for sampling from the predictive distribution of S^* automatically generate samples from the predictive distribution of S. Hence, if predictions at data locations are required, we simply combine the sampled values of S^* and S.

For parameter estimation, we need to generate samples from the posterior distribution $[\theta, \beta|Y]$. For prediction, we also require samples from the posterior distribution $[S^*|Y]$. Diggle et al. (1998) proposed an MCMC algorithm for both tasks, based on the conditional independence structure of the generalized linear geostatistical model as shown in Figure 7.8.

7.5.2 Estimation

For inference about model parameters, S^* is irrelevant. Using the structure of the conditional independence graph restricted to the nodes θ, β, S and Y, and noting that the data Y are fixed, a single cycle of the MCMC algorithm involves first sampling from $[S|\theta, \beta, Y]$, then from $[\theta|S]$, and finally from $[\beta|S, Y]$. The second stage in the cycle can in turn be broken into a sequence of samples from the univariate conditional distributions $[S_i|S_{-i}, \theta, \beta, Y]$, where S_{-i} denotes the vector S with its ith element removed. Alternatively, the vector S can be updated in a single step, as we shall discuss in Section 7.5.4. In principle, repeating this process sufficiently many times from arbitrary starting values for θ, β and S will eventually generate samples from $[\theta, \beta, S|Y]$ and hence, by simply ignoring the sampled values of S, from the required posterior $[\theta, \beta|Y]$.

We now consider the detailed form of each of these conditional distributions. Firstly, Bayes' Theorem immediately implies that

$$[\theta|S] \propto [S|\theta][\theta], \tag{7.15}$$

and that $[\beta|S, Y] \propto [Y|\beta, S][\beta]$. The structure of the generalized linear model implies that

$$p(Y|\beta, S) = \prod_{j=1}^{n} p(Y_j|\beta, S_j), \tag{7.16}$$

from which it follows that

$$p(\beta|S, Y) \propto \{\prod_{j=1}^{n} p(Y_j|\beta, S_j)\}\pi(\beta). \tag{7.17}$$

Finally, Bayes' Theorem in conjunction with the conditional independence structure of the model gives $p(S_i|S_{-i}, \theta, \beta, Y) \propto p(Y|S, \beta)p(S_i|S_{-i}, \theta)$ and (7.16) then gives

$$p(S_i|S_{-i}, \theta, \beta, Y) = \{\prod_{j=1}^{n} p(Y_j|S_j, \beta)\}p(S_i|S_{-i}, \theta). \tag{7.18}$$

Because $S(\cdot)$ is a Gaussian process, the conditional distribution $[S|\theta]$ in equation (7.15) is multivariate Gaussian, and $p(S_i|S_{-i}, \theta)$ in equation (7.18) is therefore a univariate Gaussian density. This facilitates the alternative approach of block-updating values of S jointly. In principle, we could specify any prior distributions for θ and β in equations (7.15) and (7.17). Note also that $p(Y_j|S_j, \beta) = p(y; \mu_j)$ where $\mu_j = h^{-1}\{d_j'\beta + S(x_j)\}$ and $h(\cdot)$ is the link function of the generalized linear model.

The resulting algorithm is straightforward to implement and is a general-purpose method, which makes it suitable for incorporation into a general package. However, for any particular model more efficient algorithms could certainly be devised; for further discussion, see Section 7.5.4. The general-purpose algorithm can be described more explicitly as follows. Each step uses a version of a class of methods known as Metropolis-Hastings algorithms, after Metropolis, Rosenbluth, Rosenbluth, Teller and Teller (1953) and Hastings (1970). These algorithms involve sampling a proposed update and accepting or rejecting the update with a probability which is chosen so as to guarantee convergence of the chain to the required equilibrium distribution

- Step 0. Choose initial values for θ, β and S. The initial values for θ and β should be compatible with their respective priors. Sensible initial values for S are obtained by equating each Y_i to its conditional expectation μ_i given β and $S(x_i)$, and solving for $S_i = S(x_i)$.

- Step 1. Update all the components of the parameter vector θ:

  (i) choose a new proposed value θ' by sampling uniformly from the parameter space specified by the prior;

(ii) accept θ' with probability $\Delta(\theta, \theta') = \min\left\{\frac{p(S|\theta')}{p(S|\theta)}, 1\right\}$, otherwise leave θ unchanged.

- Step 2. Update the signals, S:

  (i) choose a new proposed value, S_i', for the ith component of S from the univariate Gaussian conditional probability density $p(S_i'|S_{-i}, \theta)$, where S_{-i} denotes S with its ith element removed;

  (ii) accept S_i' with probability $\Delta(S_i, S_i') = \min\left\{\frac{p(y_i|s_i',\beta)}{p(y_i|s_i,\beta)}, 1\right\}$, otherwise leave S_i unchanged;

  (iii) repeat (i) and (ii) for all $i = 1, \ldots, n$.

- Step 3. Update all the elements of the regression parameter β:

  (i) choose a new proposed value β' from a conditional density $p(\beta'|\beta)$;
  (ii) accept β' with probability

$$\Delta(\beta, \beta') = \min\left\{\frac{\prod_{j=1}^{n} p(y_j|s_j, \beta')\, p(\beta|\beta')}{\prod_{j=1}^{n} p(y_j|s_j, \beta)\, p(\beta', |\beta)}, 1\right\},$$

otherwise leave β unchanged.

In this context, the conditional densities $p(S_i|S_{-i})$ in step 2, and $p(\beta'|\beta)$ in step 3 are called *transition kernels*. Any kernel gives a valid algorithm, but the choice can have a major impact on computational efficiency. Note that in step 2, the transition kernel is the modelled conditional distribution of S_i given S_{-i}, which seems a natural choice, whereas in step 3 the transition kernel $p(\beta'|\beta)$ is essentially arbitrary. In general, a good choice of transition kernel is problem-specific, and in our experience involves considerable trial-and-error experimentation to achieve good results.

Steps 1–3 are repeated until the chain is judged to have reached its equilibrium distribution, the so-called "burn-in" of the algorithm. Further cycling over steps 1–3 yields a sample from the posterior distribution, $[\theta, S, \beta|Y]$, which can then be processed as in Chapter 6, using properties of the empirical sample as approximations to the corresponding properties of the posterior. In principle, these samples can be made arbitrarily precise by increasing the length of the simulation run. However, in contrast to the direct Monte Carlo methods used in Chapter 6, the MCMC algorithm generates dependent samples, often in practice very strongly dependent, and the simple rule of doubling the simulation size to halve the Monte Carlo variance does not apply. It is common practice to thin the MCMC output by sampling only at every rth cycle of the algorithm. Increasing r has the effect of reducing the dependence between successive sampled values. This does not of course improve the statistical efficiency of a run of a given length, but it may be a sensible compromise between the very small gains in efficiency obtained by retaining the complete, strongly dependent sample and the convenience of storing a much smaller number of sampled values.

The usual way to display a posterior distribution obtained from an MCMC algorithm is as either a histogram or a smoothed non-parametric density estimate based on the sampled values after the algorithm is presumed to have converged. Because the MCMC sample is usually very large, typically many thousands, the choice of bin-width for the histogram, or band-width for a non-parametric smoother, is usually neither difficult nor critical. Nevertheless, the earlier warning that MCMC samples are not independent still holds, and it is always worth checking that splitting the MCMC sample in half gives essentially the same posterior for any quantity of interest. For univariate posteriors, we usually examine a superimposed plot of two *cumulative* empirical distribution functions calculated from the two half-samples, together with the cumulative prior distribution. Approximate equality of the two empirical distribution functions suggests that the sample size is adequate and gives some assurance (but no guarantee) that the algorithm is close to its equilibrium state. A large difference between the posterior and prior distributions confirms that the data are strongly informative of the parameter in question.

Displaying multivariate posteriors is less straightforward. In the bivariate case, standard practice appears to be to use an ordinary scatterplot, but a simple non-parametric smoother again provides an alternative. In higher dimensions, options include a scatterplot matrix display, a dynamic three-dimensional spinning scatterplot or a classical dimension-reducing method such as a transformation to principal components. Inspection of the bivariate posteriors for pairs of parameters can highlight possible problems of poorly identified combinations of parameter values.

7.5.3   Prediction

For prediction of properties of the realised signal, $S(\cdot)$, we need to re-introduce the fifth node, S^*, into the conditional independence graph of the model as in Figure 7.8. The goal is then to generate samples from the conditional distribution $[(S, S^*)|Y] = [S|Y][S^*|S, Y]$. The general prediction algorithm operates by adding to the three-step algorithm described in Section 7.5.2 the following.

- Step 4. Draw a random sample from the multivariate Gaussian distribution $[S^*|Y, \theta, \beta, S]$, where (θ, S, β) are the values generated in steps 1 to 3.

However, our model implies that S^* is conditionally independent of both Y and β, given S, and step 4 therefore reduces to direct simulation from the Gaussian distribution $[S^*|S, \theta]$. Specifically,

$$[S^*|S, \theta] \sim \mathrm{MVN}(\Sigma_{12}^T \Sigma_{11}^{-1} S, \; \Sigma_{22} - \Sigma_{12}^T \Sigma_{11}^{-1} \Sigma_{12}), \tag{7.19}$$

where $\Sigma_{11} = \mathrm{Var}(S)$, $\Sigma_{12} = \mathrm{Cov}(S, S^*)$ and $\Sigma_{22} = \mathrm{Var}(S^*)$. Note that if the MCMC sample is thinned, Step 4 is only needed when the corresponding sampled value of S is stored for future use.

Prediction of any target $T = T(S^*)$ then follows immediately, by computing $T_j = T(S_{(j)}^*) : j = 1, \ldots, m$ to give a sample of size m from the predictive distribution $[T|Y]$, as required; here, $S_{(j)}^*$ denotes the jth simulated sample from the predictive distribution of the vector S^*. For point prediction, we can approximate the minimum mean square error predictor, $\mathrm{E}[T(S^*)|y]$, by the sample mean, $\hat{T} = m^{-1} \sum_{j=1}^{m} T(S^*_{(j)})$. However, it will usually be preferable to examine the whole of the predictive distribution, as discussed in Section 7.5.2 in the context of posterior distributions for model parameters.

Whenever possible, it is desirable to replace Monte Carlo sampling by direct evaluation. For example, if it is possible to calculate $E[T(S^*)|S_{(j)}]$ directly, we would use the approximation

$$E[T(S^*)|Y] \approx m^{-1} \sum_{j=1}^{m} E[T(S^*)|S_{(j)}],$$

thereby reducing the Monte Carlo error due to simulation. This device is used within the package **geoRglm** (Christensen and Ribeiro Jr., 2002), which is specifically designed to fit the Poisson log-linear and binomial logistic-linear GLGM's.

7.5.4 Some possible improvements to the MCMC algorithm

As noted earlier, designing an MCMC algorithm may involve a compromise between generality of application and efficiency for specific problems. Also, the underlying theory for MCMC methods is still developing. With these qualifications, we now describe the particular algorithms used in the **geoRglm** package (Christensen and Ribeiro Jr., 2002) as suggested by Christensen (2001), Christensen and Waagepetersen (2002) and Diggle, Ribeiro Jr and Christensen (2003).

To simulate from $[S|y]$ we use the truncated Langevin-Hastings algorithm as in Christensen, Møller and Waagepetersen (2001) with the values of S block-updated as suggested by the results in Neal and Roberts (2006). This algorithm uses gradient information in the proposal distribution and has been found to work well in practice by comparison with a random walk Metropolis algorithm. To do this, we first define $S = \Omega^{1/2}\Gamma$ where $\Omega^{1/2}$ is a square root of $\Omega = \text{Var}[S]$, for example using a Cholesky factorisation, and $\Gamma \sim N(0, I)$. We then use an MCMC-algorithm to obtain a sample $\gamma_{(1)}, \ldots, \gamma_{(m)}$ from $[\Gamma|y]$, and pre-multiply each vector $\gamma_{(j)}$ by $\Omega^{1/2}$ to obtain a sample $s_{(1)}, \ldots, s_{(m)}$ from $[S|y]$.

All components of Γ are updated simultaneously in the Langevin-Metropolis-Hastings MCMC algorithm. The proposal distribution is a multivariate Gaussian distribution with mean $m(\gamma) = \gamma + (\delta/2)\nabla(\gamma)$ where $\nabla(\gamma) = \frac{\partial}{\partial\gamma} \log f(\gamma|y)$, and variance δI_n. For a generalised linear geostatistical model with canonical link function h, the gradient $\nabla(\gamma)$ has the following form:

$$\nabla(\gamma) = \frac{\partial}{\partial\gamma} \log f(\gamma|y) = -\gamma + (\Omega^{1/2})'\{y - h^{-1}(\eta)\}, \tag{7.20}$$

where $\eta = D'\beta + \Omega^{1/2}\gamma$ and h^{-1} is applied coordinatewise. If we modify the gradient $\nabla(\gamma)$ (by truncating, say) such that the term $\{y - h^{-1}(\eta)\}$ is bounded, the algorithm can be shown to be geometrically ergodic, and a Central Limit Theorem holds. The Central Limit Theorem, with asymptotic variance estimated by Geyer's monotone sequence estimate (Geyer, 1992), can then be used to assess the Monte Carlo error of the calculated prediction. The algorithm can be modified to handle other link functions, since the formula in (7.20) can be generalised to models with a non-canonical link function.

To choose the proposal variance δ, we tune the algorithm by running a few test sequences and choosing δ so that approximately 60% of the proposals are accepted. To avoid storing a large number of high-dimensional simulations we generally thin the sample; for example, we may choose to store only every 100th simulation.

Bayesian inference

We first consider Bayesian inference for a generalised linear geostatistical model using the Gaussian-Scaled-Inverse-χ^2 prior for $[\beta, \sigma^2]$ as defined in (7.6), holding ϕ fixed. The distribution $[S]$ is obtained by integrating $[S, \beta, \sigma^2]$ over β and σ^2, leading to an n-dimensional multivariate-t distribution, $t_{n_\sigma}(m_b, S_\sigma^2(R + DV_b D'))$. The posterior $[S|y]$ is therefore given by

$$p(s|y) \propto \prod_{i=1}^n g(y_i; h^{-1}(\eta_i))p(s) \tag{7.21}$$

where $p(s)$ is the density of $[S]$.

In order to obtain a sample $s_{(1)}, \ldots, s_{(m)}$ from this distribution we use a Langevin-Hastings algorithm where $\eta = D'm_b + S_\sigma(R + DV_b D')\Omega^{1/2}\Gamma$, where $\Omega = S_\sigma^2(R + DV_b D')$, and *a priori* $\Gamma \sim t_{n+n_\sigma}(0, I_n)$. The gradient $\nabla(\gamma)$ which determines the mean of the proposal distribution has the following form when h is the canonical link function,

$$\nabla(\gamma) = \frac{\partial}{\partial \gamma}\log f(\gamma|y) = -\gamma(n+n_\sigma)/(n_\sigma + \|\gamma\|^2) + (\Omega^{1/2})'\{y - h^{-1}(\eta)\}. \tag{7.22}$$

By using a conjugate prior for $[\beta, \sigma^2]$ we find that $[\beta, \sigma^2|s_{(j)}], j = 1, \ldots, m$ are Gaussian-Scaled-Inverse-χ^2 distributions with means and variances given by (7.7). Using this result, we can simulate from the posterior $[\beta, \sigma^2|y]$, and calculate its mean and variance.

For prediction, we use procedures similar to those described in Section 7.5.3. The only difference is that from (7.13), we see that for each simulation $j = 1, \ldots, m$, the conditional distribution $[S^*|s_{(j)}]$ is now multivariate t-distributed rather than multivariate Gaussian.

A word of caution is needed concerning the use of so-called non-informative priors for β and σ^2 in a generalised linear geostatistical model. The prior $1/\sigma^2$ for σ^2, recommended as a non-informative prior for the Bayesian linear Gaussian model in Section 7.2.1, here results in an improper posterior distribution (Natarajan and Kass, 2000), and should therefore be avoided. The same holds for a linear Gaussian model with a fixed positive measurement error, $\tau_0^2 > 0$. So far as we are aware, there is no consensus on what constitutes an appropriate default prior for a generalised linear mixed model.

We now allow for uncertainty also in ϕ, and adopt as our prior $\pi(\beta, \sigma^2, \phi) = \pi_{N\chi_{ScI}^2}(\beta, \sigma^2)\pi(\phi)$, where $\pi_{N\chi_{ScI}^2}$ is given by (7.5) and $\pi(\phi)$ is any proper prior.

When using an MCMC-algorithm updating ϕ, we need to calculate $(R(\phi) + DV_b D')^{1/2}$ for each new ϕ value, which is the most time-consuming part of the algorithm. To avoid this significant increase in computation time, we adopt a discrete prior for ϕ on a set of values covering the range of interest. This allows us to pre-compute and store $(R(\phi) + DV_bD')^{1/2}$ for each prior value of ϕ.

To simulate from $[S, \phi|y]$, after integrating out β and σ^2, we use a hybrid Metropolis-Hastings algorithm in which S and ϕ are updated sequentially. The update of S is of the same type as used earlier, with ϕ equal to the current value in the MCMC iteration. To update ϕ we use a random walk Metropolis update where the proposal distribution is a Gaussian distribution, but rounded to the nearest ϕ value in the discrete prior support. The output of this algorithm is a sample $(s_{(1)}, \phi_{(1)}), \ldots, (s_{(m)}, \phi_{(m)})$ from the distribution $[S, \phi|y]$.

The predictive distribution for S^* is given by

$$[S^*|y] = \int \int [S^*|S, \phi][S, \phi|y]dSd\phi.$$

To simulate from this predictive distribution, we simulate $s^*_{(j)} : j = 1, ..., m$ from the corresponding multivariate t-distributions $[S^*|s_{(j)}, \phi_{(j)}]$.

We may also want to introduce a nugget term into the specification of the model, replacing $S(x_i)$ by $S(x_i) + U_i$ where the U_i are mutually independent Gaussian variates with mean zero and variance τ^2. Here, in contrast to the Gaussian case, we can make a formal distinction between the U_i as a representation of micro-scale variation and the error distribution induced by the sampling mechanism, for example Poisson for count data. In some contexts, the U_i may have a more specific interpretation. For example, if a binary response were obtained from each of a number of sampling units at each of a number of locations, a binomial error distribution would be a natural choice, and the U_i and $S(x_i)$ would then represent non-spatial and spatial sources of extra-binomial variation, respectively. The inferential procedure is essentially unchanged, except that we now use a discrete joint prior $[\phi, \tau^2]$. Note, however, that enlarging the model in this way may exacerbate problems associated with poorly identified model parameters unless the sampling design includes replicated observations at coincident sampling locations.

The above description corresponds to the current version of **geoRglm** at the time of writing. Other possible improvements include the reparameterisation of the model suggested by Zhang (2002), and orthogonalising and standardising the conditional distribution $[S|y]$, as suggested by Christensen, Roberts and Skøld (2006).

7.6   Case studies in generalized linear geostatistical modelling

7.6.1   Simulated data

We first consider the simulated data shown in Figure 4.5. The model used to generate the data is a Poisson log-linear model whose true parameter values are: an intercept-only regression term, $\beta = 0.5$; a signal variance $\sigma^2 = 2$; and a Matérn correlation function with $\phi = 0.2$ and $\kappa = 1.5$.

Table 7.2. Summaries of the posterior for the simulated Poisson data: posterior means and 95% central quantile-based intervals.

parameters	true values	posterior mean	95% interval
β	0.5	0.4	[0.08 , 1.58]
σ^2	2.0	1.24	[0.8 , 2.76]
ϕ	0.2	0.48	[0.3 , 1.05]

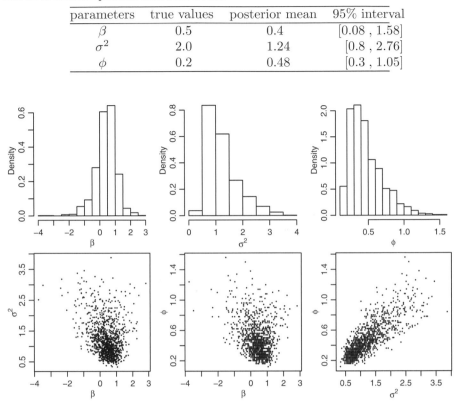

Figure 7.9. Histograms and scatterplots of the sample from the posterior distribution for model parameters for the simulated Poison data-set.

For the analysis, we adopt the correct model and treat the Matérn shape parameter as fixed, $\kappa = 1.5$. For the correlation parameter we use a uniform discrete prior with 100 support points in the interval $(0, 2)$.

After tuning the MCMC algorithm, we used a burn-in of 10,000 iterations, followed by a further 100,000 iterations, from which we stored every 100th value to give a sample of 1,000 values from the posterior and predictive distributions.

Table 7.2 shows summaries of the posterior distributions, whilst Figure 7.9 shows univariate histograms and bivariate scatterplots of the samples from the posterior. Although the central quantile-based interval for the parameter ϕ excludes the true value, the upper right-hand panel of Figure 7.9 shows that this is a consequence of the asymmetry in the posterior for ϕ.

We obtained samples from the predicted distribution at two locations $(0.75, 0.15)$ and $(0.25, 0.50)$. For the former the median of the sample from the predictive distribution is 1.79 with prediction uncertainty of 0.97, whereas for the latter the corresponding values are 4.39 and 2.32, respectively. Figure 7.10

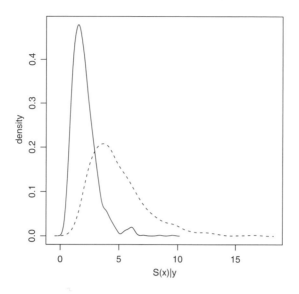

Figure 7.10. Density estimated prediction distributions at two selected locations for the Poisson simulated data.

shows density estimates for the samples from the predictive distribution at each of these locations.

7.6.2 Rongelap island

Our second case study is based on the data from Example 1.2. The data were collected as part of an investigation into the residual contamination arising from nuclear weapons testing during the 1950's. This testing programme resulted in the deposition of large amounts of radioactive fallout on the pacific island of Rongelap. The island has been uninhabited since the mid-1980's. A geostatistical analysis of residual contamination levels formed one component of a wide-ranging project undertaken to establish whether the island was safe for re-habitation. Earlier analyses of these data are reported by Diggle et al. (1997) who used log-Gaussian kriging, and by Diggle et al. (1998) who used the model-based approach reported here, but with minor differences in the detailed implementation.

For our purposes, the data consist of nett photon emission counts Y_i over time-periods t_i at locations x_i indicated by the map in Figure 1.2. The term "nett" emission count refers to the fact that an estimate of natural background radiation has been subtracted from the raw count in such a way that the datum Y_i can be attributed to the local level of radioactive caesium at or near the surface of the island. The background effect accounts for a very small fraction of the total radioactivity, and we shall ignore it from now on.

The gamma camera which records photon emissions integrates information received over a circular area centred on each location x_i. There is also a progressive "dilution" effect with increasing distance from x_i. Hence, if $\lambda^*(x)$ denotes the true rate of photon emissions per unit time at location x, the raw count at location x will follow a Poisson distribution with mean

$$\mu(x) = t(x) \int w(x - u)\lambda^*(u)du \qquad (7.23)$$

where $t(x)$ denotes the observation time corresponding to location x. The function $w(\cdot)$ decays to zero over a distance of approximately 10 metres, but we do not know its precise form. However, the minimum distance between any two locations in Figure 1.2 is 40 metres. Hence, rather than model $\lambda^*(\cdot)$ in (7.23) directly, we will model $\lambda(\cdot)$, where

$$\lambda(x) = \int w(x - u)\lambda^*(u)du.$$

Our general objective is to describe the spatial variation in the spatial process $\lambda(x)$. Note that any spatial correlation in $\lambda(\cdot)$ induced by the integration of the underlying process $\lambda^*(\cdot)$ operates at a scale too small to be identified from the observed data. Hence, any empirically observed spatial correlation must be the result of genuine spatial variation in local levels of residual contamination, rather than an artefact of the sampling procedure.

The sampling design for the Rongelap island survey was a lattice plus in-fill design of the kind which we discuss in Chapter 8. This consists of a primary lattice overlaid by in-fill squares in selected lattice cells. In fact, the survey was conducted in two stages. The primary lattice, at 200m spacing, was used for the first visit to the island. The in-fill squares were added in a second visit, to enable better estimation of the small-scale spatial variation. For the second-stage sample, two of the primary grid squares were selected randomly at either end of the island. As we discuss in Chapter 8, inclusion of pairs of closely spaced points in the sampling design can be important for identification of spatial covariance structure, and therefore for effective spatial prediction when the true model is unknown. In this application, we can also use the in-fill squares to make an admittedly incomplete assessment of the stationarity of the underlying signal process. For example, if we let y denote the nett count per second, then for 50 sample locations at the western end of the island including the two in-fill squares, the sample mean and standard deviation of $\log(y)$ are 2.17 and 0.29, whilst for 53 locations covering the eastern in-fill area the corresponding figures are 1.85 and 0.35. Hence, the western extremity is the more heavily contaminated, but the variation over two areas of comparable size is quite similar; see also the exercises at the end of this chapter.

Taking all of the above into consideration, we adopt a Poisson log-linear model with log-observation time as an offset and a latent stationary spatial process $S(\cdot)$ in the linear predictor. Explicitly, if Y_i denotes the nett count over observation time t_i at location x_i, then our modelling assumptions are

- conditional on a latent spatial process $S(\cdot)$, the Y_i are mutually independent Poisson variates with respective means μ_i, where

$$\log \mu_i = \log t_i + \beta + S(x_i); \qquad (7.24)$$

- $S(\cdot)$ is a stationary Gaussian process.

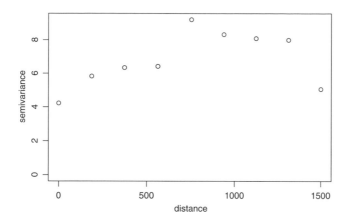

Figure 7.11. Empirical variogram of the transformed Rongelap data.

In order to make a preliminary assessment of the covariance structure, we transform each datum (Y_i, t_i) to $Y_i^* = \log(Y_i/t_i)$. Under the assumed log-linear structure of the proposed model, we can think of Y_i^* as a noisy version of the unobserved $S(x_i)$. Hence, the sample variogram of the observed values of Y_i^* should give a qualitative pointer to the covariance structure of the latent process $S(\cdot)$. Figure 7.11 shows the resulting empirical variogram. The relatively large intercept suggests that measurement error, which in the model derives from the approximate Poisson sampling distribution of the nett counts, accounts for a substantial proportion of the total variation. However, there is also clear structure to the empirical variogram, indicating that the residual spatial variation is also important. The convex shape of the empirical variogram suggests that this spatial variation is fairly rough in character.

The remaining results in this section are taken from the analysis reported in Diggle et al. (1998), who used the powered exponential family (3.7) to model the correlation structure of $S(\cdot)$,

$$\rho(u) = \exp\{-(u/\phi)^\kappa\}.$$

Recall that for this model, $\kappa \le 2$ and unless $k = 2$ the model corresponds to a mean-square continuous but non-differentiable process $S(\cdot)$. Also, for $\kappa \le 1$ the correlation function $\rho(\cdot)$ is convex, which would be consistent with our earlier comment on the shape of the sample variogram of the Y_i^*.

The priors for β, σ^2, ϕ and κ were independent uniforms, with respective ranges $(-3, 7)$, $(0, 15)$, $(0, 120)$ and $(0.1, 1.95)$. The corresponding marginal posteriors have means $1.7, 0.89, 22.8$ and 0.7, and modes $1.7, 0.65, 4.7$ and 0.7. Note in particular the strong positive skewness in the posterior for ϕ, and confirmation that the data favour $\kappa < 1$ i.e., a convex correlation function for the process $S(\cdot)$.

For prediction, Diggle et al. (1998) ran their MCMC algorithm for 51,000 iterations, discarded the first 1000 and then sampled every 100 iterations to give a sample of 500 values from the posterior distributions of the model parameter, and from the predictive distribution of the surface $S(x)$ at 960 locations forming

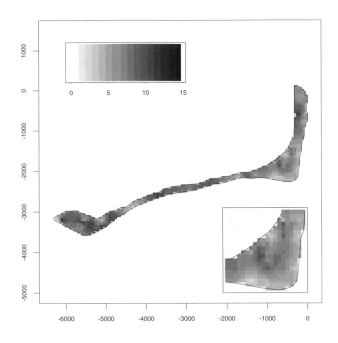

Figure 7.12. Point predictions of intensity (mean count per second) for the Rongelap data. Each value is the mean of a Monte Carlo sample of size 500.

a square lattice to cover the island at a spacing of 50 metres. By transforming each sampled $S(x)$ to $\lambda(x) = \exp\{\beta + S(x)\}$, they obtained a sample of 500 values from the predictive distribution of the spatially varying intensity, or mean emission count per second, over the island. Figure 7.12 shows the resulting point-wise mean surface. This map includes the southeast corner of the island as an enlarged inset, to show the nature of the predicted small-scale spatial variation in intensity. Note also the generally higher levels of the predictions at the western end of the island.

A question of particular practical importance in this example is the pattern of occurrence of relatively high levels of residual contamination. Maps of point predictions like the one shown in Figure 7.12 do not give a very satisfactory answer to questions of this kind because they do not convey predictive uncertainty. One way round this is to define a specific target T and to show the whole of the predictive distribution of T, rather than just a summary. To illustrate this approach, the left-hand panel of Figure 7.13 shows the predictive distribution of $T = \max\{\lambda(x)\}$, where the maximum is computed over the same 960 prediction locations as were used to construct Figure 7.12. Note that the predictive distribution extends far beyond the maximum of the point-wise predictions of $\lambda(x)$ shown in Figure 7.12. The two versions of the predictive distribution refer to predictions with and without allowance for parameter uncertainty, showing that for this highly non-linear functional of $S(\cdot)$ parameter uncertainty makes

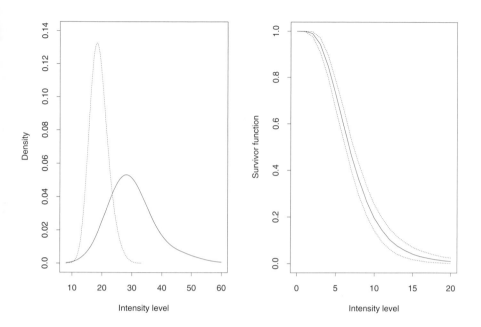

Figure 7.13. Predictive inference for the Rongelap island data. The left-hand panel shows the predictive distribution of $T = \max\lambda(x)$, computed from a grid to cover the island at a spacing of 50 metres, with (solid line) and without (dashed line) allowance for parameter uncertainty. The right-hand panel shows point predictions (solid line) and 95% credible limits (dashed lines) for $T(z)$, the areal proportion of the island for which intensity exceeds z counts per second.

a material difference. Of course, by the same token so does the assumed parametric model which underlies this predictive inference. By contrast, predictive inferences for the point-wise values of $\lambda(x)$ are much less sensitive to parameter uncertainty.

The right-hand panel of Figure 7.13 summarises the predictive distributions of a family of targets, $T(z)$ equal to the areal proportion of the island for which $\lambda(x) > z$. The predictive distribution for each value of z is summarised by its point prediction and associated 95% central quantile-based interval.

7.6.3 Childhood malaria in The Gambia

Our third case study uses the data from Example 1.3. This concerns spatial variation in the prevalence of malarial parasites in blood samples taken from children in village communities in The Gambia, Africa. Figure 1.3 shows a map of the village locations. Note that these represent only a small fraction of the village communities in The Gambia. The strongly clustered arrangement of sampled villages is clearly not ideal from a theoretical point of view, but reflects the practical necessities of field work in difficult conditions on a limited budget. Thomson, Connor, D'Alessandro, Rowlingson, Diggle, Cresswell and Greenwood (1999) describe the background to the study in more detail. The analysis described here was previously reported by Diggle, Moyeed, Rowlingson and Thomson (2002).

Two similarities between this and the Rongelap data are the following. Firstly, there is a natural sampling model for the responses conditional on the underlying signal, in this case the binomial distribution. Secondly, there is no natural way to specify a mechanistic model for the residual spatial variation, hence we again adopt a stationary Gaussian process as a flexible empirical model. One notable difference from the Rongelap case study is that covariates are recorded both at village-level and at individual child-level. Village-level covariates are a satellite-derived measure of the green-ness of the surrounding vegetation, which is a predictor of how favourable the local environment is for mosquitos to breed, and a binary indicator of whether the village had its own health centre. Child-level covariates are sex, age and bed-net use. Sex is thought to be unimportant. Age is almost certainly important because of chronic infections. Bed-net use is also likely to be important; this is a three-level factor coded as a pair of binary covariates, one for bed-net use itself, the other indicating whether the net was treated with insecticide.

In this example, valid inferences about covariate effects are probably more important in practice than estimation of spatial variation in its own right. The primary role of the latent spatial process $S(\cdot)$ is to guard against spuriously significant covariate effects which might result from ignoring the spatial correlation inherent in the data. Residual spatial effects, estimated after adjustment for covariate effects, are nevertheless of some interest, since identifying areas of unusually high or low residual prevalence might point to other, as yet unidentified, risk factors.

In order to take account of child-level covariates, the natural statistical model for these data is a generalized linear mixed model for the binary outcome observed on each child. Diggle, Moyeed, Rowlingson and Thomson (2002) initially tried to fit a model which included both spatial and non-spatial random effects, as follows. Let p_{ij} denote the probability that the jth child in the ith village gives a positive blood-test result. Then,

$$\log\{p_{ij}/(1 - p_{ij})\} = \alpha + \beta' z_{ij} + U_i + S(x_i). \tag{7.25}$$

where the U_i are mutually independent Gaussian random effects with mean zero and variance τ^2, whilst $S(x)$ is a zero-mean stationary Gaussian process with variance σ^2 and correlation function $\rho(u) = \exp\{-(|u|/\phi)^\kappa\}$. Diggle, Moyeed, Rowlingson and Thomson (2002) were unable to estimate jointly the two variance components τ^2 and σ^2. This suggests that the random effect part of model (7.25) is over-ambitious. Because the scientific focus in this example is on estimating covariate effects, it would be tempting to eliminate the spatial effect, $S(x)$ altogether. The non-spatial random effects U_i would then represent a classical generalized linear mixed model with the simplest possible random effects structure, which could be fitted either through approximate likelihood-based methods or, more pragmatically, by fitting a classical generalised linear

model with a simple adjustment factor applied to the nominal standard errors
to take account of extra-binomial variation at the village-level; see, for example,
chapter 4 of McCullagh and Nelder (1989). Note, however, that the implicit esti-
mands in these two analyses would differ. Specifically, the β parameters in (7.25)
measure covariate effects conditional on village-level random effects, whilst the
classical generalised linear model estimates different parameters, β^* say, which
measure covariate effects averaged over the distribution of the random effects.
In general, $|\beta^*| < |\beta|$ element-wise, as discussed, for example, in chapter 7 of
Diggle, Heagerty, Liang and Zeger (2002).

Diggle, Moyeed, Rowlingson and Thomson (2002) report that in the non-
spatial version of (7.25), the predicted village-level random effects $\hat{U}_i$ showed
substantial spatial correlation. Since spatial variation, although not the primary
focus of the analysis, is of some interest, they therefore persevered with a spatial model, but omitted the non-spatial random effects U_i. They also made a
pragmatic modification to the stationarity assumption for $S(x)$, in response to
the strongly clustered nature of the sampling design, by introducing a five-level
factor corresponding to villages included in each of the five separate surveys
from which the data were assembled. The five areas corresponded to villages in
the western, central and eastern parts of The Gambia, but with the western and
eastern parts further divided into villages north and south of the River Gambia.

The model was fitted using an MCMC algorithm as described in Diggle,
Moyeed, Rowlingson and Thomson (2002). Table 7.3 summarises the results in
terms of marginal posterior means and 95% credible intervals for the model
parameters. With regard to the spatial covariance parameters, the widths of
the credible intervals underline the difficulty of estimating these parameters
precisely, reinforcing our earlier comments that the inferences are potentially
sensitive to prior specifications. We would argue, however, that our formal approach to inference simply reveals difficulties which are hidden when more *ad
hoc* methods of parameter estimation are used.

7.6.4   Loa loa prevalence in equatorial Africa

Our final case study again relates to binomial sampling for estimating tropical disease prevalence. However, in contrast to The Gambia malaria example,
the spatial variation in prevalence has direct policy implications. The analysis
reported here is taken from Diggle, Thomson, Christensen, Rowlingson, Obsomer, Gardon, Wanji, Takougang, Enyong, Kamgno, Remme, Boussinesq and
Molyneux (2006).

Predicting the spatial distribution of *Loa loa* prevalence is important because
it affects the operation of the African Programme for Onchocerciasis Control
(APOC), a major international programme to combat onchocerciasis in the
wet tropics. APOC oversees the mass treatment of susceptible communities
with the drug ivermectin, which is effective in protecting against onchocerciasis, but has been observed to produce severe, and occasionally fatal, reactions in
some individuals who are heavily co-infected with *Loa loa* parasites. Boussinesq,
Gardon, Kamgno, Pion, Gardon-Wendel and Chippaux (2001) confirmed empir-

Table 7.3. Point estimates (posterior means and medians) and 95% central quantile-based credible intervals for the parameters of the model fitted to The Gambia malaria data.

Parameters	95% credible interval		Mean	Median
α	-2.9665	2.6243	-0.1312	-0.0780
β_1 (age)	0.0005	0.0009	0.0007	0.0007
β_2 (untreated)	-0.6731	-0.0420	-0.3578	-0.3594
β_3 (treated)	-0.7538	0.0884	-0.3295	-0.3259
β_4 (green-ness	-0.0857	0.0479	-0.0201	-0.0208
β_5 (PHC)	-0.7879	0.1299	-0.3448	-0.3499
β_6 (area 2)	-1.1442	0.5102	-0.3247	-0.3316
β_7 (area 3)	-1.4086	0.5586	-0.5321	-0.5592
β_8 (area 4)	-0.1095	2.4253	1.0494	1.0170
β_9 (area 5)	0.1648	2.6063	1.3096	1.3251
σ^2	0.3118	1.0502	0.5856	0.5535
ϕ	0.9158	10.2007	2.5223	1.4230
δ	0.0795	2.7846	1.0841	0.9374

ically that such individuals are more likely to be found in areas with high local prevalence. They also investigated the relationship between local prevalence and the proportion of individuals infected at levels sufficiently high to render them apparently susceptible to severe reactions. Informed by this and other work, current APOC policy is to put in place precautionary measures before mass distribution of ivermectin in communities for which the local prevalence of *Loa loa* is thought to exceed 20%. However, direct estimation of prevalence throughout the relevant area, which covers most of the wet tropical zone, is impractical. One response to this is to conduct a geostatistical analysis of survey data on *Loa loa* prevalence obtained from scattered village communities, and to map the predictive probability that the 20% policy intervention threshold is exceeded.

The data for the analysis include empirical prevalence estimates obtained by microscopic examination for the presence of *Loa loa* parasites in blood samples taken from a total of 21,938 individuals from 168 villages. The locations of these villages are shown in Figure 7.17. To these data are added environmental explanatory variables measured on a regular grid covering the study region at a spacing of approximately 1 km. The environmental variables are those chosen by Thomson, Obsomer, Kamgno, Gardon, Wanji, Takougang, Enyong, Remme, Molyneux and Boussinesq (2004) in a non-spatial logistic regression analysis of the prevalence data. These are elevation, obtained from the United States Geological Survey website http://edcdaac.usgs.gov/gtopo30/hydro/africa.html, and a vegetation index (Normalised Difference Vegetation Index, NDVI), a measure of the greenness of the vegetation which is derived from satellite data available from the web-site http://free.vgt.vito.be.

Let Y_i denote the number of positive samples out of n_i individuals tested at location x_i. Diggle et al. (2006) fitted a generalised linear geostatistical model

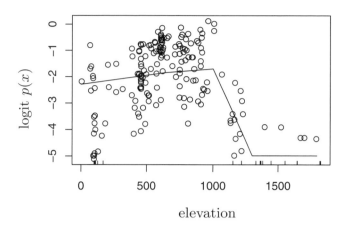

Figure 7.14. Piece-wise linear function used in the spatial model to describe the effect of elevation on *Loa loa* prevalence.

in which the Y_i are assumed to be conditionally independent binomial variates given an unobserved Gaussian process $S(x)$. The mean response at x_i is $E[Y_i] = n_i p(x_i)$, where $p(x_i)$ depends on the values at x_i of the chosen environmental variables and on $S(x_i)$. Specifically, the model assumes that

$$\log(p(x)/\{1-p(x)\}) = \beta_0 + f_1(\text{ELEV}) + f_2(\max(\text{NDVI})) + f_3(SD(\text{NDVI})) + S(x).$$
$$(7.26)$$

In (7.26), the functions $f_1(\cdot)$, $f_2(\cdot)$ and $f_3(\cdot)$ are piece-wise linear functions which capture the effects of elevation and NDVI on *Loa loa* prevalence at the location x. Only linear functions $f_j(\cdot)$ were considered initially, but exploratory analysis showed threshold effects in the impact of both elevation and NDVI on prevalence, which were confirmed as qualitatively reasonable on substantive grounds; for example, the biting fly which is the vector for *Loa loa* transmission is known not to survive at high elevations. The rationale for including both the maximum and standard deviation of NDVI, each calculated for each grid location from repeated satellite scans over time, is that together they capture, albeit crudely, the effects of overall greenness of the local vegetation and seasonal variation in local greenness, both of which were thought to affect the ability of the *Loa loa* vector to breed successfully.

Figures 7.14, 7.15 and 7.16 show the construction of the piece-wise linear functions $f_1(\cdot)$, $f_2(\cdot)$ and $f_3(\cdot)$ through which the model represents the effects of elevation and NDVI on *Loa loa* prevalence in the spatial model (7.26). There is a positive association between elevation and prevalence up to a threshold of 1000 metres above sea-level, beyond which prevalence drops sharply, becoming effectively zero above 1300 metres. The effect of maximum NDVI on prevalence is modelled as a linear increase up to a value of 0.8, and constant thereafter;

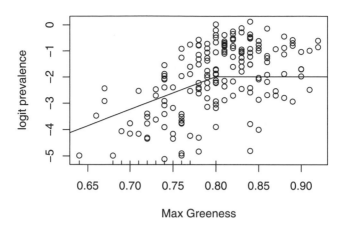

Figure 7.15. Piece-wise linear function used in the spatial model to describe the effect of maximum NDVI on *Loa loa* prevalence.

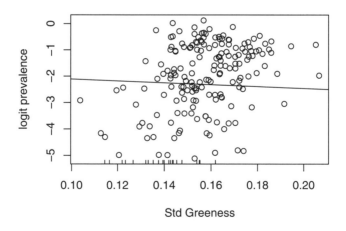

Figure 7.16. Piece-wise linear function used in the spatial model to describe the effect of standard deviation of NDVI on *Loa loa* prevalence.

the defined range of NDVI is from zero to one. Finally, standard deviation of NDVI shows a very weak association with prevalence which we represent as a simple linear effect.

The model for $S(\cdot)$ is a stationary Gaussian process with mean zero and covariance function

$$\gamma(u) = \sigma^2 \{\exp(-u/\phi) + \nu^2 I(u = 0)\},$$

where $I(\cdot)$ denotes the indicator function. The form of the covariance structure was chosen after inspecting the empirical variogram of residuals from a non-spatial binomial logistic regression model. The nugget term, $\tau^2 = \sigma^2 \nu^2$, represents non-spatial extra-binomial variation in village-level prevalence. This is attributed to non-specific social and demographic attributes of individual village communities which are not spatially dependent.

Prior specifications for the model parameters were as follows: for (β, σ^2), an improper prior $\pi(\beta, \sigma^2) \propto 1$; for the correlation parameter, ϕ, a proper uniform prior, $\pi(\phi) = c^{-1} : 0 \leq \phi \leq c$, with $c = 1$ degree of latitude/longitude at the equator, or approximately 100 km; for the relative nugget parameter, ν^2, a fixed value $\nu^2 = 0.4$. The upper limit for ϕ and the fixed value of ν^2 were again chosen after inspection of the residual empirical variogram. Fixing ν^2 was a pragmatic strategy to circumvent problems with poor identifiability.

The target for prediction is the predictive probability, for any location x, that $p(x)$ exceeds 0.2 given the data. Monte Carlo methods were used to construct a map of these predictive probabilities on a regular grid with spacing approximately 1 km, chosen to match the spatial resolution of the explanatory variables in the model.

The inferential procedure divides naturally into two steps. The first step is to generate samples from the joint posterior distribution of the model parameters (β, σ^2, ϕ) and the spatial random effects S at the village locations, using a Markov chain Monte Carlo (MCMC) algorithm as described in Section 7.5.4. The second step is then to generate samples from the predictive distribution of the spatial random effects at all locations in the square grid of prediction locations; this step requires only direct simulation from a multivariate Gaussian distribution.

Table 7.4 gives summaries of the posterior distributions for the model parameters, based on sampling every 1000th iteration from 1,000,000 iterations of the MCMC algorithm. The correspondence between the β parameters and (7.26) is as follows: β_0 is the intercept; β_1, β_2 and β_3 are the slope parameters in the linear spline for the elevation effect ($f_1(\text{ELEV})$ in (7.26)), covering the elevation ranges 0–650 metres, 650–1000 metres and 1000–1300 metres, respectively; the linear spline for the effect of maximum NDVI has slope β_4 between 0.0 and 0.8, and is constant thereafter ($f_2(\max(\text{NDVI}))$ in equation (7.26)); β_5 is the slope of the linear effect of the standard deviation of NDVI ($f_3(\text{SD}(\text{NDVI}))$ in equation (7.26)).

For prediction of the 20% prevalence contour at location x_0, say, the following Monte Carlo approximation was used,

$$P[p(x_0) > 0.2 | y] \approx (1/m) \sum_{j=1}^{m} P(S(x_0) > c | S_j, \beta_j, \sigma_j^2, \phi_j),$$

Table 7.4. Posterior means and standard deviations for parameters of the model fitted to the *Loa loa* data. See text for detailed explanation.

parameter	Mean	Std. dev.
β_0	−11.38	2.15
β_1	0.0007	0.0007
β_2	0.0004	0.0011
β_3	−0.0109	0.0016
β_4	12.45	2.92
β_5	−3.53	4.77
σ^2	0.58	0.11
ϕ^2	0.70	0.18

where, in addition to previous notation, $c = \log(0.2/(1 - 0.2))$ and a subscript j indicates the jth of m samples from the posterior distribution of the model parameters. As noted earlier, inference was based on the empirical posterior/predictive distributions of every 1000th sample from 1,000,000 MCMC iterations, hence $m = 1000$. Note also that $S(x_0)$ conditional on S, β, σ^2 and ϕ follows a Gaussian distribution function, yielding an explicit expression for $P(S(x_0) > t|S, \beta, \sigma^2, \phi)$ in (7.27).

The total number of prediction locations in the grid at 1 km spacing was 550,000, albeit including a small proportion of off-shore locations for which predictions are formally available but of no relevance. To ease the computational load of the predictive calculation, the prediction locations were divided into sub-sets consisting of approximately 10,000 locations. Separate predictions were then made within each sub-set and combined to produce the maps shown here.

The map of estimated prevalence obtained from the spatial model is presented in Figure 7.17. This map corresponds broadly with the map obtained by Thomson et al. (2004) using a non-spatial logistic regression model, but its local features differ because of the effect of the residual spatial term, $S(x)$.

Figure 7.18 shows the predictive probability map obtained from the spatial model. Dark grey areas are those where there is a relatively high predictive probability that the policy intervention threshold of 20% is exceeded. Likewise, pale grey areas are those where the predictive probability of exceeding the 20% threshold is relatively small. The intermediate areas can be considered as areas of high uncertainty. The maps are clearer in the colour versions which are posted on the book's website. There, red-brown areas correspond to predictive probabilities greater than 0.7, orange-yellow areas to predictive probabilities less than 0.3, whilst predictive probabilities between 0.3 and 0.7 are coded in shades of pink.

As we would expect, there is a qualitative similarity between the maps in Figures 7.17 and 7.18 but the quantitative differences are sufficient materially to affect their interpretation. Note in particular that the relationship between the value of the empirical prevalence at a particular location x and the corresponding predictive probability of exceeding the 20% threshold involves an

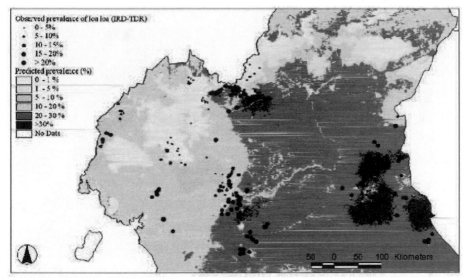

Figure 7.17. Point estimates of *Loa loa* prevalence. Village locations which provided survey data are shown as solid dots, coded by size to correspond to the observed proportions of positive blood-test results amongst sampled individuals. The spatially continuous grey-scale map shows, at each point on the map, the point prediction of the underlying prevalence.

interplay between the influences of the environmental explanatory variables at x and of the empirical prevalences at nearby locations.

From the point of view of the people who need to make local decisions, the obvious limitation of the predictive probability map is the high degree of uncertainty in many parts of the study region. The solution is to obtain more survey data, concentrating on the areas of high uncertainty. However, the kind of data used to construct the map are expensive to collect and additional sampling on the scale required is unlikely to be affordable. In response to this impasse, Takougang, Wanji, Yenshu, Aripko, Lamlenn, Eka, Enyong, Meli, Kale and Remme (2002) have developed a simple questionnaire instrument, RAPLOA, for estimating local prevalence, as a low-cost alternative to the parasitological sampling method used here. Combining the data from parasitological and RAPLOA sampling is a problem in bivariate generalised linear modelling, which takes us beyond the scope of this book but is the subject of work-in-progress by Ciprian Crainiceanu, Barry Rowlingson and Peter Diggle.

7.7 Computation

7.7.1 Gaussian models

For plug-in prediction based on the Gaussian linear model, our computational implementation comprises two steps: parameter estimation, for instance using `likfit()`; and point prediction using `krige.conv()` Within the Bayesian

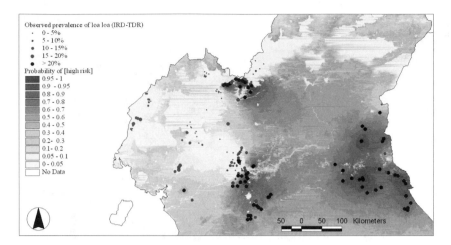

Figure 7.18. Predictive inference for the *Loa loa* data. Village locations which provided survey data are shown as solid dots, coded by size and colour to correspond to the observed proportions of positive blood-test results amongst sampled individuals. The spatially continuous map shows, at each point on the map, the predictive probability that the underlying prevalence is greater than 20%.

framework this distinction is much less clear and we therefore implement inference as a whole in a single function, `krige.bayes()`. This function can return either or both posterior distributions for model parameters and predictive distributions for the unobserved signal $S(x)$ at prediction locations x.

The **geoR** function `krige.bayes()` implements Bayesian inference for the Gaussian linear model whereas the functions `binom.krige.bayes()` and `pois.krige.bayes()` in the package **geoRglm** implement methods for the binomial and Poisson generalised linear geostatistical models. Our initial examples in this section illustrate some of the options available to the user of `krige.bayes()`

```
> args(krige.bayes)

function (geodata, coords = geodata$coords, data = geodata$data,
    locations = "no", borders, model, prior, output)
NULL
```

The function requires arguments specifying the data object, model and prior information. Implementation is made more convenient by using the supplementary functions `model.control()` and `prior.control()` to specify the model and prior information. Specification of the prediction locations is optional; if these are omitted, `krige.bayes()` carries out Bayesian inference about the model parameters but does not perform any spatial prediction. Other optional arguments include the border of the study area, which is particularly relevant for non-rectangular areas, and options to specify which results should be included in the output. For the latter, summaries for the predictive distribution can be obtained using `output.control()`, as discussed in relation to the `krige.conv()` function in Chapter 6. Additional options relevant to Bayesian inference, namely the number of samples to be drawn from the posterior and predictive distributions, are specified by the arguments `n.posterior` and `n.predictive`, respectively. The remaining arguments define the required summaries of the predictive distributions, again as discussed in relation to the `krige.conv()` function.

```
> args(output.control)
```

```
function (n.posterior, n.predictive, moments, n.back.moments,
    simulations.predictive, mean.var, quantile, threshold, sim.means,
    sim.vars, signal, messages)
NULL
```

We now illustrate the use of these functions for Bayesian inference in the Gaussian linear model by showing the sequence of commands for the Bayesian analysis of the elevation data, as reported in Section 7.4.1.

Firstly, a call to `model.control()` specifies the option of fitting a first degree polynomial trend. This option can be used for covariates which correspond to the coordinates or to another covariate. Therefore, we specify the trend on the data coordinates using the argument `trend.d` and on the prediction locations in `trend.l`.

Next, we set the support points for the default independent, discrete uniform priors for ϕ and $\nu^2 = \tau^2/\sigma^2$. For σ^2 and β we use the default prior $\pi(\beta, \sigma^2) = 1/\sigma^2$. Using the function `output.control()` we define the number of samples to be drawn from the predictive distributions and ask for their analytically computed mean and variance. Note that all of the control functions have other arguments with default values; as always, the calls to the functions only specify values for the arguments for which we want to override the defaults.

```
> MC <- model.control(trend.d = "1st", trend.l = "1st",
+     kappa = 1.5)
> PC <- prior.control(phi.discrete = seq(0, 6, l = 21),
+     phi.prior = "reciprocal", tausq.rel.prior = "unif",
+     tausq.rel.discrete = seq(0, 1, l = 11))
> OC <- output.control(n.post = 1000, moments = T)
```

After setting the control functions, we proceed to the computations required for Bayesian inference. We first define a grid of prediction points to cover the study area, then call the function `krige.bayes()`, passing the results of the control functions as arguments.

```
> set.seed(268)
> skb <- krige.bayes(elevation, loc = locs, model = MC,
+     prior = PC, output = OC)
```

The resulting object is of the class `krige.bayes`. An object of this class has two main elements, `posterior` and `predictive`, which are used to store samples and other information concerning the predictive and posterior distributions. The `krige.bayes` class also includes methods for the generic functions `image()`, persp() and contour. These operate in a similar way as for krige.conv(), to facilitate displaying spatial predictions. The generic summary() function can be used to summarise the results with regard to spatial prediction. Finally, the command plot(skb) generates Figure 7.2, showing the marginal posteriors for the parameters ϕ and ν.

7.7.2   Non-Gaussian models

In this section we show how we used the **geoRglm** package to fit a Poisson log-linear model to the simulated data whose Bayesian analysis we presented in Section 7.6.1.

The algorithm uses the Langevin-Hastings algorithm to simulate from the predictive distribution of the random effect S at each of the data locations; we tuned the algorithm by adjusting the proposal distribution so as to achieve an acceptance rate of about 60%. For the correlation parameter ϕ, we used a random walk proposal, $\phi' = \phi + Z$, where Z has mean zero and variance v^2, and adjusted v^2 to achieve an acceptance rate around 25 to 30%. The tuning phase involved the following commands, adjusting the arguments S.scale and phi.sc until the quoted approximate acceptance rates were obtained.

```
> set.seed(371)
> MCc <- mcmc.control(S.scale = 0.014, n.iter = 5000,
+       thin = 100, phi.sc = 0.15)
> PGC <- prior.glm.control(phi.prior = "uniform",
+       phi.discrete = seq(0, 2, by = 0.02), tausq.rel = 0)
> pkb <- pois.krige.bayes(dt, prior = PGC, mcmc = MCc)
```

After tuning the algorithm, the full-length run was initiated by the following commands.

```
> set.seed(371)
> MCc <- mcmc.control(S.scale = 0.025, phi.sc = 0.1, n.iter = 110000
+       burn.in = 10000, thin = 100, phi.start = 0.2)
> PGC <- prior.glm.control(phi.prior = "exponential", phi = 0.2,
+       phi.discrete = seq(0, 2, by = 0.02), tausq.rel = 0)
> OC <- output.glm.control(sim.pred = T)
> locs <- cbind(c(0.75, 0.15), c(0.25, 0.5))
> pkb <- pois.krige.bayes(dt, loc = locs, prior = PGC,
+       mcmc = MCc, out = OC)
```

For the binomial model with logit link functions, the steps are essentially the same, except that the **geoRglm** function used is binom.krige.bayes() instead of pois.krige.bayes.

7.8   Exercises

7.1. Consider the stationary Gaussian model in which $Y_i = \beta + S(x_i) + Z_i$: $i = 1, \ldots, n$, where $S(x)$ is a stationary Gaussian process with mean zero, variance σ^2 and correlation function $\rho(u)$, whilst the Z_i are mutually independent $N(0, \tau^2)$ random variables. Assume that all parameters except β are known. Derive the Bayesian predictive distribution of $S(x)$ for an arbitrary location x when β is assigned an improper uniform prior, $\pi(\beta)$ constant for all real β. Compare the result with the ordinary kriging formulae given in Chapter 6.

7.2. Repeat the calculations of exercise 7.1, but assigning a proper Gaussian prior, $\beta \sim N(m, v)$. Explore how varying m and v affects the predictions obtained for the following, one-dimensional synthetic data, taking $\sigma^2 = 1$, $\tau^2 = 0.25$ and $\rho(u) = \exp(-u/5)$.

x_i	1.00	2.00	3.00	4.00	5.00	6.00	7.00	8.00	9.00
y_i	5.44	5.40	4.44	4.04	4.19	4.94	4.94	5.71	5.63

x_i	10.00	11.00	12.00	13.00	14.00	15.00	16.00	17.00	18.00
y_i	6.09	5.95	5.08	5.64	5.75	4.51	4.98	5.30	5.82

x_i	19.00	20.00	21.00	22.00	23.00	24.00	25.00
y_i	5.11	5.60	5.45	5.15	5.88	5.60	5.33

7.3. Let $S \sim N(\mu, \sigma^2)$, and suppose that, conditional on S, random variables $Y_i : i = 1, \ldots, n$ are mutually independent, identically distributed, $Y_i \sim N(S, 1)$. Find the predictive distribution of S given $Y = (Y_1, \ldots, Y_n)$ when

(a) it is known that $\mu = 10$ and $\sigma^2 = 1$;
(b) it is known that $\sigma^2 = 1$, but μ is unknown and is assigned a prior distribution, $\mu \sim N(10, v^2)$.

Compare the predictive distributions obtained under (a) and (b) for various combinations of n and v^2, and comment generally.

7.4. Use the Rongelap data to obtain two sub-sets of data, from the western and eastern ends of the island, and in each case include the data from the two in-fill squares. Define a response variable $y = \log(\text{number of emissions per second})$. Compute and compare sample variograms for the two sub-sets. Suggest a parametric model for these data and use (classical or Bayesian) likelihood-based methods to investigate whether a good fit can be obtained for both sets of data using a common set of parameters. Discuss the implications for the analysis of the complete Rongelap data.

7.5. Experiment with your own simulated data from the Poisson log-linear model and investigate the sensitivity of the MCMC algorithm to different choices for the model parameters and for the tuning parameters.

7.6. Reproduce the simulated binomial data shown in Figure 4.6. Use **geoRglm** in conjunction with priors of your choice to obtain predictive distributions for the signal $S(x)S$ at locations $x = (0.6, 0.6)$ and $x = (0.9, 0.5)$.

7.7. Compare the predictive inferences which you obtained in Exercise 7.6 with those obtained by fitting a linear Gaussian model to the empirical logit transformed data, $\log\{(y + 0.5)/(n - y + 0.5)\}$.

7.8. Compare the results of Exercises 7.7 and 7.8 and comment generally.

8
Geostatistical design

In this chapter, we consider the specific design problem of where to locate the sample points $x_i : i = 1, ..., n$. In particular applications other design issues, such as what to measure at each location, what covariates to record and so forth, may be at least as important as the location of the sample points. But questions of this kind can only be addressed in specific contexts, whereas the sample-location problem can be treated generically.

In Chapter 1 we introduced the terms *non-uniform*, meaning that the method of constructing the design incorporates systematic variation in the sampling intensity over the study region, and *preferential*, meaning that the point process which determines the sample locations and the signal process $S(x)$ are stochastically dependent. In this chapter, we shall consider both uniform and non-uniform designs, but will restrict our attention to non-preferential designs. As noted earlier, geostatistical analyses typically assume, if only implicitly, that a non-preferential design has been used. A valid analysis of data obtained from a preferential design requires the more general theoretical framework of marked point processes, as discussed in Section 4.4.

In some applications, the design is essentially unrestricted, in the sense that any point in the study region is a potential sample point. It will then usually be appropriate to consider only uniform designs unless we have prior knowledge that the character of the spatial signal, $S(x)$, varies systematically over the study region. For example, if it were known that $S(x)$ was essentially constant in particular sub-regions, there should be no need to sample intensively in those sub-regions.

In other applications, the choice of sample points may be restricted in some way. One form of restriction is when the study region includes sub-regions which are of interest for prediction but inaccessible for sampling. An example would be the assessment of contaminated land in urban areas when it is required to predict the pattern of soil contamination over the whole of a potential redevelopment site, but the site includes derelict buildings where soil sampling is impossible (Van Groenigen and Stein, 1998; Van Groenigen, Siderius and Stein, 1999; Van Groenigen, Pieters and Stein, 2000). A second kind of restriction is when there are only a finite number of candidate sample points. An example would be spatial prediction using sample points chosen from an existing monitoring network, which may originally have been established for some other purpose. The two kinds of restriction are often combined when sampling in an urban environment and installation of the sampling equipment requires a particular kind of location, such as a flat-roofed building.

All too frequently in our experience, the sampling design is presented as a *fait accompli*. When this is the case, it is always worth asking why the particular design has been used before proceeding with any formal analysis.

Different designs will be optimal for different purposes. In the geostatistical setting, a particularly relevant contrast is between designs which are efficient for parameter estimation, and designs which are efficient for spatial prediction. For either purpose, an optimal design will typically depend not only on the chosen optimality criterion but also on the underlying model parameters, which are almost invariably unknown. This has led some authors to propose model-independent design criteria, defined in terms of the geometry of the sample locations. For example, Royle and Nychka (1988) consider minimising a measure of the average distance between sample locations and locations at which predictions are required. Either or both of the sample and prediction locations may represent locations of particular scientific interest or, more pragmatically, a fine grid to cover the whole study area. The resulting designs tend to be spatially regular in appearance. Royle and Nychka (1988) give an example in which the sample locations are to be selected as a sub-set of an existing network of environmental monitors in Chicago, and the prediction locations form a regular grid over the whole city.

Amongst model-dependent approaches to geostatistical design, we can contrast those which focus on parameter estimation, and those which focus on spatial prediction under an assumed model. In the first category, Russo (1984), Warrick and Myers (1987), Zimmerman and Homer (1991) and Muller and Zimmerman (1999) consider the design problem from the perspective of variogram estimation. As we have argued in earlier chapters, we are sceptical of treating variogram estimation as a primary objective for formal inference. In the second category, McBratney, Webster and Burgess (1981), McBratney and Webster (1981), Winkels and Stein (1997), Spruill and Candela (1990), and Ben-jamma, Marino and Loaiciga (1995) all considered the design problem using the maximum or average prediction variance over the study region as the design criterion.

Several authors have used computationally sophisticated Monte Carlo algorithms, such as simulated annealing, to search for optimal designs without any prior restrictions on their geometry. Examples include Van Groenigen and

Stein (1998), Van Groenigen et al. (1999), Van Groenigen et al. (2000) and Lark (2002).

In the remainder of this chapter we take a model-based view of the design problem. We discuss briefly the choice of study region within which all of the sampling locations must lie. We then focus on the problem of choosing sample locations within a specified study region. We compare and contrast designs which are efficient for prediction or for parameter estimation under an assumed model, before using the Bayesian paradigm to obtain designs which are efficient for prediction whilst allowing for uncertainty in the underlying model parameters. Our general aim is to provide some insight into the kinds of design which are easily implemented and reasonably efficient under a wide range of conditions, rather than to search for a strictly optimal design for any particular problem. In particular, unless the design points must be chosen from a pre-existing set of locations, we favour designs with a modified lattice structure. The more sophisticated simulated annealing approach typically results in designs which are irregular, but which are otherwise similar in character to the kinds of design we advocate, in the sense that they exhibit a degree of spatial regularity combined with some closely spaced sub-sets of locations.

8.1   Choosing the study region

The study region, A say, within which sample points x_i will be confined, is often pre-determined by the context of the investigation; for example, the whole of Rongelap island was the only natural choice of study region for the data of Example 1.2. When A is not pre-determined, we need to choose its size and its shape. For example, in ecological applications the potential study area may be an entire habitat, only a small proportion of which will be selected for detailed study.

In theory, the shape of A is relatively unimportant. In practice, a long, thin shape limits the scope to investigate directional effects and from this point of view a compact shape such as a square or circular A is preferable.

The size of A is more important, in that it limits the range of spatial scales which can be investigated. If A is too small, the full range of spatial variation in the underlying measurement process may not be captured. Empirical evidence for this would be provided by a sample variogram which fails to level out at large distances. At the other extreme, if A is too large then, given a limitation on total resources, the individual sample points will tend to be widely separated and small-scale spatial effects may go undetected. One way to compromise between these competing considerations is to sample from several widely separated sub-areas. The results from the different sub-areas can then be pooled if the analysis of the subsequent data indicates that this is justified.

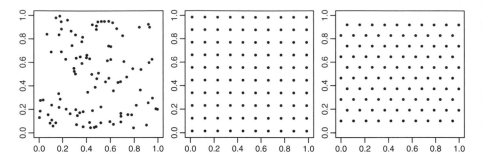

Figure 8.1. Three designs with $n = 100$ locations: random (left-hand panel), square lattice (centre panel) and triangular lattice (right-hand panel).

8.2 Choosing the sample locations: uniform designs

In practice, geostatistical designs are often chosen informally, rather than by the use of explicit design criteria. In this section, we show examples of four simple classes of design which we shall use later in the chapter to illustrate the impact of the choice of design on the subsequent inferences which can be made from the data.

In a *completely random* design, the design points $x_i : i = 1, ..., n$ form an independent random sample from the uniform distribution on A. The left-hand panel of Figure 8.1 gives an example with $n = 100$. A completely random design guarantees that the design is independent of the underlying spatial phenomenon of interest, $S(x)$, which is a requirement for validity of standard geostatistical methods of inference. However, from a spatial perspective this design is potentially inefficient because it can lead to a very uneven coverage of A. At the opposite extreme, a regular lattice design achieves even coverage of A and retains the requirement of being independent provided the position of the first lattice point is chosen independently of $S(x)$. The centre and right-hand panels of Figure 8.1 show two examples with $n = 100 = 10 \times 10$. It is arguable that a triangular lattice, rather than a square lattice, represents the extreme of spatial regularity, but the convenience of laying out the orthogonal rows and columns of a square lattice seems to prevail in practice.

From the perspective of spatial prediction, a lattice design is usually more efficient than a completely random design. When, as is typical, $S(x)$ has positive-valued correlation structure, close pairs of points are wasteful because they provide little more information about $S(x)$ than does a single point, and the lattice design automatically excludes such close pairs. However, this ignores two important practical considerations. Firstly, if the data include a substantial nugget variance, replicate measurements at close, or even identical, locations do convey useful additional information. Secondly, and more generally, close pairs of point are often especially helpful in the estimation of parameters which define the covariance structure of the model. These considerations suggest that some compromise between spatial regularity, for even coverage of A, and closely spaced points, for estimation of unknown model parameters, might be desirable.

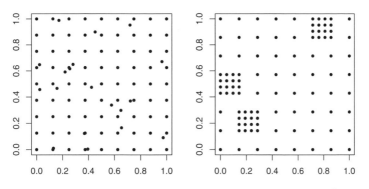

Figure 8.2. Two designs with 100 locations: lattice with close pairs (left-hand panel) and lattice plus in-fill (right-hand panel).

Two classes of design which formalise this idea are *lattice plus close pairs* and *lattice plus in-fill* designs.

The *lattice plus close pairs* design consists of locations in a regular $k \times k$ lattice at spacing Δ together with a further m points, each of which is located uniformly at random within a disc of radius $\delta = \alpha\Delta$ whose centre is at a randomly selected lattice location. We use the notation $(k \times k, m, \alpha)$, noting that from the design point of view, the choice of the distance scale is arbitrary and hence Δ is irrelevant. The left-hand panel of Figure 8.2 shows a $(9 \times 9, 19, 0.05)$ lattice plus close pairs design on the unit square.

The *lattice plus in-fill* design consists of locations in a regular $k \times k$ lattice at spacing Δ together with further locations in a more finely spaced lattice within m randomly chosen cells of the primary lattice. We use the notation $(k \times k, m, r \times r)$, where each in-filled lattice cell consists of an $r \times r$ lattice and therefore involves $r^2 - 4$ additional locations. The right-hand panel of Figure 8.2 shows an $(8 \times 8, 3, 4 \times 4)$ lattice plus in-fill design on the unit square.

8.3   Designing for efficient prediction

Suppose that the study area A has been chosen. In this section we assume, for convenience, that A is a square of unit side-length, but this choice is not critical to our conclusions.

Our objective is to choose n sample locations x_i, where n is fixed by resource constraints, so as to obtain the "best" predictions of the underlying signal process $S(x)$. Recall that for any target T, we define the mean square prediction error as $\mathrm{MSE}(T) = \mathrm{E}[(\hat{T} - T)^2]$. When $T = S(x)$ for a particular location x, we write $M(x) = \mathrm{E}[\{\hat{S}(x) - S(x)\}^2]$. Amongst many possible criteria to define "best" we consider the following:

1. minimise the maximum of $M(x)$ over all $x \in A$, where $M(x) = \mathrm{MSE}\{S(x)\}$;

2. minimise the spatial average of $M(x)$,

$$\int_A M(x)dx$$

3. minimise MSE(T) where T is the spatial average of $S(x)$,

$$T = \int_A S(x)dx$$

.

For criteria of this kind, intuition suggests that spatially regular designs will perform well. As noted earlier, model-free design criteria of the kind proposed by Royle and Nychka (1988) tend in practice to produce spatially regular designs when the goal is to optimise some version of average predictive performance over the whole study region.

Early work in the forestry literature, summarised in Matérn (1986, chapter 5) confirms the intuitively sensible idea that regular lattice designs are generally efficient for prediction of the spatial average of $S(x)$. The same intuition suggests that regular lattice designs should be efficient whenever the optimality criterion is neutral with regard to location, in the sense that all parts of the study region A are of equal scientific interest. The problem of estimating a spatial average is related to the classical survey sampling problem of estimating the mean of a finite population, and whether this is better approached through design-based or model-based methods. See, for example, Bellhouse (1977) or Sarndal (1978).

We now give some numerical comparisons between two contrasting designs on the unit square: a regular $k \times k$ square lattice; and a completely random spatial distribution of $n = k^2$ locations. An example of each of the two designs when $k = 10$ was shown in Figure 8.1. For each of these two designs, we have evaluated each of the three design criteria listed above, using a 25 by 25 square lattice of prediction locations as a discrete approximation to the whole of A, and generating data from replicated simulations of the stationary Gaussian model with mean μ, signal variance σ^2, correlation function $\rho(u) = \exp(-u/\phi)$ and nugget variance τ^2. For the simulations, we fixed $\mu = 0$, $\sigma^2 + \tau^2 = 1$ but varied ϕ and the noise-to-signal variance ratio τ^2/σ^2.

Table 8.1 summarises the results of the simulation experiment. The lattice design dominates the random design, in the sense that in all cases it produces a smaller value for the design criterion. However, the relative efficiency depends on both the design criterion and the model parameters.

8.4   Designing for efficient parameter estimation

When the design objective is to estimate model parameters efficiently, we need to balance two competing considerations. On the one hand, pairs of sample points which are spatially close relative to the range of the spatial correlation are needed to identify correlation parameters. On the other hand, the measurements from spatially close points are themselves correlated, and therefore less informative about marginal parameters (in particular, mean and variance) than

Table 8.1. Comparison of random and square lattice designs, each with $n = 100$ sample locations, with respect to three design criteria: spatial maximum of mean square prediction error $M(x)$; spatial average of mean square prediction error $M(x)$; scaled mean square error, $100 \times MSE(T)$, for $T = \int S(x)dx$. The simulation model is a stationary Gaussian process with parameters $\mu = 0$, $\sigma^2 + \tau^2 = 1$, correlation function $\rho(u) = \exp(-u/\phi)$ and nugget variance τ^2. The tabulated figures are averages of each design criterion over $N = 500$ replicate simulations.

		max $M(x)$		average $M(x)$		$MSE(T)$	
Model parameters		Random	Lattice	Random	Lattice	Random	Lattice
$\tau^2 = 0$	$\phi = 0.05$	9.28	8.20	0.77	0.71	0.53	0.40
	$\phi = 0.15$	5.41	3.61	0.40	0.30	0.49	0.18
	$\phi = 0.25$	3.67	2.17	0.26	0.19	0.34	0.10
$\tau^2 = 0.1$	$\phi = 0.05$	9.57	8.53	0.81	0.76	0.54	0.41
	$\phi = 0.15$	6.22	4.59	0.50	0.41	0.56	0.28
	$\phi = 0.25$	4.44	3.34	0.37	0.30	0.47	0.22
$\tau^2 = 0.3$	$\phi = 0.05$	10.10	9.62	0.88	0.86	0.51	0.40
	$\phi = 0.15$	7.45	6.63	0.65	0.60	0.68	0.43
	$\phi = 0.25$	6.23	5.70	0.55	0.51	0.58	0.38

would be the case with more widely spaced points. The wider the class of models under consideration, the stronger the case for including close pairs of points in the design. If the functional form of the theoretical correlation function is not tightly constrained by the assumed model, then we would need empirical correlation estimates over a wide range of distances in order to ascertain its shape. For estimating a single unknown correlation parameter a more restricted range of distances, such as would be obtained using a regular lattice design, may be adequate.

Other general considerations in designing for parameter estimation are the following. Firstly, as we have seen in Chapter 6, the nugget variance has a big effect on how spatial prediction operates in practice, and it is therefore particularly important to estimate this parameter accurately. This supports the inclusion of at least some close pairs of points in the design, as exemplified by either the lattice plus close pairs or the lattice plus in-fill design. Secondly, and as is typical of most branches of applied statistics, we do not know in advance the values of the model parameters, or indeed the precise form of the model itself. Hence, designs which perform reasonably over a wide range of parameter values, and which help diagnose lack-of-fit to particular models, may be preferable to ones which are optimal for particular values of the model parameters.

8.5   A Bayesian design criterion

When efficient prediction is the goal, the comparative results reported in Section 8.3 can be criticised for ignoring the effects of parameter uncertainty. In the Bayesian setting, the predictive distribution of $S(\cdot)$ is a weighted average of plug-in predictive distributions,

$$[S|Y] = \int [S|Y,\theta][\theta|Y]d\theta, \tag{8.1}$$

where θ denotes the complete set of model parameters. Hence, a design which optimises a suitable property of the plug-in predictive distribution, for example the average prediction variance over a set of target locations, may not be optimal from the Bayesian viewpoint if it results in a highly dispersed posterior distribution for θ. By optimising with respect to the Bayesian predictive distribution (8.1), we achieve the desired objective whilst allowing for the effects of parameter uncertainty.

This suggests that Bayesian-optimal designs for spatial prediction will tend to be spatially less regular than plug-in-optimal designs, because they need to compromise between the regular designs which are efficient for prediction under a known model, and the less regular designs which are efficient for parameter estimation. Formally, this compromise is realised through (8.1), which penalises inefficient parameter estimation in a natural way. Diggle and Lophaven (2006) give explicit results for two different settings, which they call retrospective and prospective design. The first is where, because of resource constraints, an existing design is to be modified by deletion of one or more points and data are available from the existing design. The second is where a design is to be chosen in advance of any data collection. The remainder of this section draws heavily on the material in Diggle and Lophaven (2006).

8.5.1   Retrospective design

To motivate the retrospective design problem, Diggle and Lophaven (2006) considered the following problem in environmental monitoring. In order to measure the spatial variation in an environmental variable of concern, a relatively dense network of monitoring locations is established, and data are collected from each. Armed with the data from this initial network, and seeking to economise on its continued maintenance, we wish to reduce the number of sites whilst incurring the least possible associated loss of environmental information. As a specific, but hypothetical example, suppose that an existing network consists of the 50 sites whose spatial distribution is shown in Figure 8.3, and that we wish to reduce this network to one with 20 sites.

The data on the initial design were generated using the stationary Gaussian model with mean $\mu = 0$, signal variance $\sigma^2 = 1$ and exponential correlation function $\rho(u) = \exp(-u/\phi)$ with parameter $\phi = 0.3$. For the noise-to-signal variance ratio, $\nu^2 = \tau^2/\sigma^2$, we used each of the values $\nu^2 = 0$, 0.3 and 0.6.

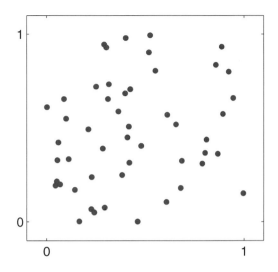

Figure 8.3. The locations in the initial design of a hypothetical monitoring network with 50 sites (from Diggle and Lophaven, 2006).

Denote by $v(x)$ the prediction variance, $\mathrm{Var}\{S(x)|Y\}$, at the point x. For the examples in this section, Diggle and Lophaven (2006) used the design criterion

$$\bar{v} = \int_A v(x)dx, \qquad (8.2)$$

which they approximate by the average of $v(x)$ over points x in a regular 6×6 grid. For the classical design approach, ignoring parameter uncertainty, the prediction variance was evaluated using ordinary kriging for prediction. For the Bayesian approach the prior for ϕ was uniform on $(0, 2.35)$, whilst for $(\mu, \sigma^2|\phi)$ a diffuse prior proportional to $1/\sigma^2$ was used. For the ratio ν^2 Diggle and Lophaven (2006) compared results obtained by assuming known ν^2 and by assigning to ν^2 a prior uniform distribution on $(0, 1)$. Posteriors were computed by direct simulation as described in Chapter 7.

Figure 8.4 shows the final designs of 20 locations according to the three different treatments of parameter uncertainty, for each of the three considered values of ν^2. The most striking feature of the results is that the classical design criterion, which ignores parameter uncertainty, leads to spatially regular designs with well-separated monitoring sites, whereas either variant of the Bayesian approach leads to retention of some close pairs of sites, representing the previously noted compromise between designing for prediction and for parameter estimation. Within the Bayesian approach, treating ν^2 as known or unknown generally led to comparable degrees of spatial regularity in the selected designs.

Example 8.1. Salinity monitoring in the Kattegat basin
As an application of their retrospective design criterion, Diggle and Lophaven (2006) considered the deletion of points from an existing network established

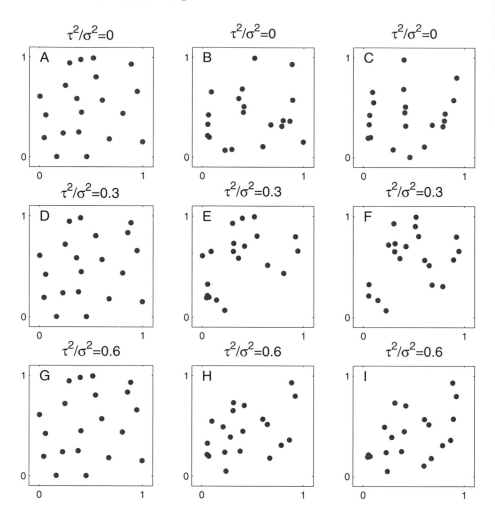

Figure 8.4. Final designs obtained when a hypothetical monitoring network is reduced from 50 to 20 sites. The designs in panels A, D and G assume that all model parameters are known, and correspond to noise-to-signal variance ratios $\nu^2 = \tau^2/\sigma^2 = 0, 0.3, 0.6$ respectively. The designs in panels B, E and H assume that all model parameters except ν^2 are unknown, whilst those in panels C, F and I assume that all parameters are unknown, again corresponding to true noise-to-signal variance ratios $\nu^2 = 0, 0.3, 0.6$.

to monitor spatial variation in salinity within the Kattegat basin, between Denmark and Sweden. The initial network consists of 70 sites, whose spatial distribution is shown in Figure 8.5. Each measurement represents average salinity over a particular time period.

The data showed a north-south trend in salinity, and for the design evaluations Diggle and Lophaven (2006) used a Gaussian model with a linear trend surface, $\mu(x) = \beta_0 + \beta_1 x_1 + \beta_2 x_2$ where x_1 and x_2 denote the east-west and north-south coordinates of a generic location x. The ratio $\nu^2 = \tau^2/\sigma^2$ was fixed

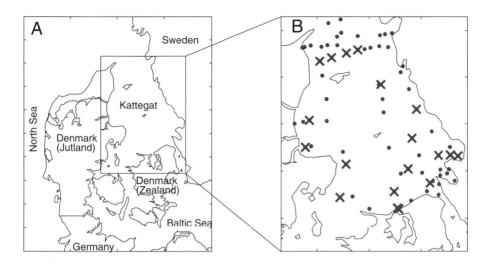

Figure 8.5. Locations of the 70 monitoring stations ($\bullet$ and $\times$) measuring salinity in the Kattegat basin, and of the 20 stations ($\times$) which are retained in the final design.

at its estimated value, $\nu^2 = 0.42$, partly to economise on computation but also because, as noted above, fixing ν^2 seems generally to make a relatively small difference to the chosen design. For ϕ, a uniform prior on the interval from 10 to 100 kilometres was used, for $(\beta, \sigma^2|\phi)$ a diffuse prior proportional to $1/\sigma^2$. The design criterion was the spatial average of the prediction variance, as given by (8.2), but approximated by averaging $v(x)$ over 95 locations in a regular grid covering the Kattegat area at a spacing of 15 kilometres. The resulting network of 20 retained monitoring stations is shown in Figure 8.5. It consists mostly of well-separated stations, but with some close pairs, again illustrating how the Bayesian approach compromises between designing for prediction and for estimation.

8.5.2 Prospective design

In its purest form, the prospective design problem is to locate a given number of points, $x_1, ..., x_n$ say, within a designated planar region A without any prior constraints. Modified versions of the problem limit the x_i to be chosen either from designated sub-regions of A or from within a finite set of candidate locations. In practice, the first two versions of the problem can be approximated by the third version, by defining candidate locations as a fine grid to cover those parts of A which are available for sampling.

In the retrospective design problem, the data from the existing design were used to estimate model parameters, θ, and the resulting posterior for θ was used to evaluate the prediction variance, $v(x) = \text{Var}\{S(x)|Y\}$, which would result from a modified design. When designing prospectively, we do not have any data Y. Diggle and Lophaven (2006) therefore proceeded by simulating data under an assumed value for θ and used as design criterion the expectation of the spatially averaged prediction variance $E[\bar{v}]$ with respect to the distribution of Y at the true parameter value θ_0. Hence, the design criterion is

$$E[\bar{v}] = \int_A E_{Y|\theta_0}[v(x)]dx. \tag{8.3}$$

Evaluation of (8.3) using Monte Carlo methods proceeds as follows. First simulate s independent data-sets, $Y_k : k = 1, ..., s$ from the model with parameter values $\theta_0 = (\mu_0, \sigma_0^2, \phi_0, \tau_0^2)$. From each simulated data-set, calculate the corresponding value, $\bar{v}_k$, of the spatially averaged prediction variance. Finally, use the sample average of the $\bar{v}_k$ over the s simulations as an approximation to $E[\bar{v}]$.

Example 8.2. Comparing regular lattice, lattice plus close pairs and lattice plus in-fill designs.

This example compares the regular 8×8 lattice with the $(7 \times 7, 15, 0.5)$ lattice plus close pairs design and the $(7, 3, 3 \times 3)$ lattice plus in-fill design. All designs were constructed on a unit square region, with the lattice spacing adjusted accordingly.

Although designs of the same type vary because of the random selection of the secondary locations, Diggle and Lophaven (2006) found that this had only a small impact on the spatially averaged prediction variance, and therefore used only five independent replicates of each design to evaluate the expectation of the average prediction variance. The model used in each case was the linear Gaussian model with constant mean μ, signal variance σ^2, nugget variance τ^2 and exponential correlation function, $\rho(u; \phi) = \exp(-|u|/\phi)$, with prior specifications as follows; for ϕ, a uniform prior on $(0, 1.3)$; for the ratio $\nu^2 = \tau^2/\sigma^2$, a uniform prior on $(0, 1)$ and for $(\mu, \sigma^2|\phi, \nu^2)$, a diffuse prior proportional to $1/\sigma^2$.

Evaluations of the chosen design criterion for each of the three candidate designs with true parameter values $\beta = 0$, $\sigma^2 = 1$, $\phi = 0.2, 0.4, 0.6, 0.8, 1.0$ and $\nu^2 = 0.0, 0.2, 0.4, 0.6, 0.8$ are summarised in Figure 8.6. This shows that the lattice plus close pairs design results in lower values of the design criterion compared to both the lattice plus in-fill design and the regular 8×8 lattice, meaning that predictions are computed more accurately from the lattice plus close pairs design. In contrast, the performance of the lattice plus in-fill design is only slightly better than that of the regular lattice.

Although a single example cannot be definitive, the qualitative message which we take from Example 8.2 is that, whilst adding some closely spaced groups of points to a lattice design is beneficial, the lattice plus in-fill design risks committing too high a proportion of the total sampling effort to the closely spaced points. For example, in the Rongelap island study of Example 1.2, the in-fills account for 100 of the 157, or more than two-thirds, of the points in the design. Some justification in this application is provided by the fact that the data were initially collected using only the 57 locations in the primary lattice, so re-considering the primary lattice spacing was not an option. Also, in terms of effort in the field, locating and taking measurements from a 5 by 5 in-fill was easier than locating several in-fills with fewer points in each.

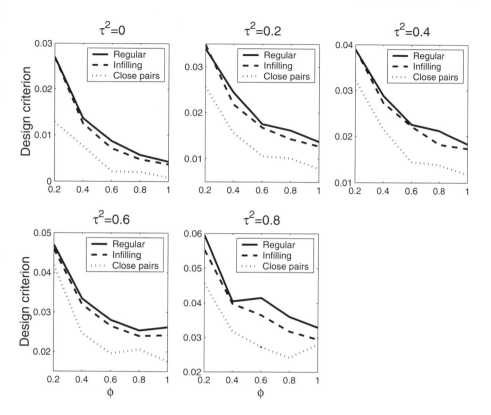

Figure 8.6. Prospective design results showing the difference in efficiency of the regular 8×8 lattice, the $(7 \times 7, 15, 0.5)$ lattice plus close pairs design, and the $(7 \times 7, 3, 3 \times 3)$ lattice plus in-fill design.

8.6 Exercises

8.1. Consider a stationary Gaussian process in one spatial dimension, in which the design consists of n equally spaced locations along the unit interval with $x_i = (-1 + 2i)/(2n) : i = 1, ..., n$. Suppose that the process has unknown mean μ but known variance $\sigma^2 = 1$ and correlation function $\rho(u) = \exp(-u/\phi)$ with known $\phi = 0.2$.

Investigate, using simulation if necessary, the impact of n on the efficiency of the maximum likelihood estimator for μ. Does the variance of $\hat{\mu}$ approach zero in the limit as $n \to \infty$? If not, why not?

8.2. Repeat exercise 8.1, but treating each of σ^2 and ρ in turn as the unknown parameter to be estimated.

8.3. Repeat exercises 8.1 and 8.2, but now considering the design to consist of n equally spaced locations $x_i = i : 1, ..., n$.

8.4. Discuss the similarities and differences amongst your results from exercises 8.1, 8.2 and 8.3.

8.5. An existing design on the unit square A consists of four locations, one at each corner of A. Suppose that the underlying model is a stationary Gaussian process with mean μ, signal variance σ^2, correlation function $\rho(u) = \exp(-u/\phi)$ and nugget variance τ^2. Suppose also that the objective is to add a fifth location, x, to the design in order to predict the spatial average of the signal process $S(x)$ with the smallest possible prediction mean square error, assuming that the model parameter values are known.

(a) Guess the optimal location for the fifth point.
(b) Suppose that we use the naive predictor $\bar{y}$. Compare the mean square prediction errors for the original four-point and the augmented five-point design.
(c) Repeat, but using the simple kriging predictor.

Appendix A
Statistical background

A.1 Statistical models

In very general terms, a statistical *model* specifies the form of the distribution of a vector-valued random variable Y in terms of a vector-valued *parameter*, θ. We write the model as $[Y|\theta]$, in which the square bracket notation means "the distribution of" and the vertical bar denotes conditioning, hence $[Y|\theta]$ means "the distribution of Y for a given value of θ." The essence of a parameter is that its value is unknown. However, if we can observe data y which can be assumed to form a realisation of Y, then a central objective of statistical *inference* is to use the specified model to find out as much as possible about θ. For any statistical model $[Y|\theta]$, the *likelihood*, $\ell(\theta)$, is algebraically equal to the joint probability density function of Y, but considered as a function of θ, rather than of Y. The likelihood function is fundamental to both classical and Bayesian statistical inference; where these two schools of inference differ is in how they interpret and use the likelihood function.

A.2 Classical inference

In classical inference, $\ell(\theta)$ is considered as a function of the non-random variable θ, with Y held fixed at its observed value y. It is usually more convenient to work with the log-likelihood, $L(\theta) = \log \ell(\theta)$. Values of θ which correspond to relatively large or small values of $L(\theta)$ are considered to be more or less supported by the evidence provided by the data, y. Thus, for a point estimate of θ we use the *maximum likelihood estimate*, $\hat{\theta}$, defined to be the value which maximises $L(\theta)$. Similarly, for an interval estimate of θ we use a *likelihood inter-*

val, defined to be the set of values for which $L(\theta) \geq L(\hat{\theta}) - c$, for some suitable value of c.

The log-likelihood is a function of the observed data y, and is therefore a realisation of a random variable whose distribution is induced by that of Y. To emphasise this, we shall temporarily use the expanded notation $L(\theta, y)$ for the observed log-likelihood and $L(\theta, Y)$ for the corresponding random variable. By the same token, we write $\hat{\theta}(y)$ for an observed value of $\hat{\theta}$ and $\hat{\theta}(Y)$ for the corresponding random variable. The derivatives of the log-likelihood function with respect to elements of θ play an important role in classical inference. In particular, we define the *information matrix*, $I(\theta)$ to have (j, k)th element

$$I_{jk} = \mathrm{E}_Y \left[-\frac{\partial^2}{\partial \theta_j \partial \theta_k} L(\theta, Y) \right].$$

Then, the properties of likelihood-based inference are summarised by the following two theorems.

Theorem A.1. $\hat{\theta}(Y) \sim \mathrm{MVN}(\theta, I(\theta)^{-1})$

Theorem A.2. $2\{L(\hat{\theta}(Y), Y) - L(\theta, Y)\} \sim \chi_p^2$, where p is the dimensionality of θ.

Both theorems are asymptotic in n, the dimensionality of y, and hold under very general, but not universal, conditions. The main exceptions in practice are when the true value of θ is on a boundary of the parameter space (for example, a zero component of variance) or when one or more elements of θ define the range of Y; for example, if the Y_i are uniformly distributed on $(0, \theta)$. For detailed discussion, see for example Cox and Hinkley (1974).

In both of these theorems, and in classical inference more generally, the status of θ is that there is a true value of θ which is fixed, but unknown and unknowable (the literal meaning of "parameter" is "beyond measurement"). Inferences about θ take either of two forms. In *hypothesis testing*, we hypothesise a particular value, or a restricted set of values, for θ and ask, using either of our two theorems, whether the data are reasonably consistent with the hypothesised value. In *parameter estimation*, we assemble all hypothesised values with which the data are reasonably consistent into a *confidence set* for θ.

The formal meaning of a hypothesis test, with *significance level* α, is the following. We divide the space of all possible data-sets y into a *critical region*, $\mathcal{C}$, and its complement in such a way that *if the hypothesis under test is true*, then under repeated sampling of the data y from the underlying model, y will fall within $\mathcal{C}$ with probability p. Should this actually occur, we then *reject* the hypothesis at significance level p. Conventionally, $\alpha = 0.05$ or smaller. Thus, in rejecting a hypothesis we are, in effect, saying that either an event of small probability has been observed or the hypothesis under test is false, with an implicit invitation to the reader to conclude the latter.

The formal meaning of a confidence set, with *confidence level* β, is that it is a random set constructed in such a way that, over repeated sampling of the data y from the underlying model, the confidence set will contain the true, fixed but unknown value of θ with probability β. It is tempting, but incorrect, to interpret this as meaning that β is the probability that θ is contained in the actual confidence set obtained from the observed data. In classical inference, a parameter is not a random variable and probabilities cannot be ascribed to it. Conventionally, confidence levels are set at $\beta = 0.95$ or larger.

There is a close duality between hypothesis testing and the evaluation of a confidence set. Specifically, any procedure for testing the hypothesis $\theta = \theta_0$ can be converted to a procedure for evaluating a confidence set for θ; a $\beta = 1 - \alpha$ confidence set consists of all hypothesised values θ_0 which are *not* rejected using a test with significance level α.

Operationally, likelihood-based inference is simple when θ is low dimensional, the log-likelihood function is easily evaluated and the conditions for the validity of theorems A.1 and A.2 are satisfied. The key numerical task is one of maximisation with respect to different possible values of θ. Specifically, using $\hat{\theta}$ to denote the maximum likelihood estimator, define the *deviance function* for θ to be

$$D(\theta) = 2\{L(\hat{\theta}) - L(\theta)\},$$

and write $c_p(\beta)$ for the β-quantile of the χ_p^2 distribution i.e., $P\{\chi_p^2 \leq c_p(\beta)\} = \beta$. Then, using theorem A.2:

- the set of all values of θ such that $D(\theta) \leq c_p(\beta)$ is, asymptotically, a β-level confidence set for θ;

- if $D(\theta_0) > c_p(\beta)$, the hypothesis $\theta = \theta_0$ is rejected at the $\alpha = 1 - \beta$ level of significance.

A very useful extension to likelihood-based inference is the method of *profile likelihood*, which operates as follows. Suppose that θ is partitioned as $\theta = (\theta_1, \theta_2)$, with corresponding numbers of elements p_1 and p_2. Suppose also that our primary objective is inference about θ_1. For each possible value of θ_1, let $\hat{\theta}_2(\theta_1)$ be the value of θ_2 which maximises the log-likelihood with θ_1 held fixed. We call $L_P(\theta_1) = L\{\theta_1, \hat{\theta}_2(\theta_1)\}$ the *profile log-likelihood* for θ_1. Then, an extension to theorem A.2 states that we can treat the profile log-likelihood as if it were a log-likelihood for a model with parameter θ_1 of dimension p_1. Specifically, if we define the deviance function for θ_1 as

$$D(\theta_1) = 2\{L_P(\hat{\theta}_1) - L_P(\theta_1)\}$$

then, asymptotically, $D(\theta_1)$ is distributed as chi-squared on p_1 degrees of freedom. This result provides a method for eliminating the effects of the nuisance parameters θ_2 when making inference about the parameters of interest, θ_1. Note once more that the key mathematical operation is one of maximisation.

A.3   Bayesian inference

In Bayesian inference, the likelihood again plays a fundamental role, and θ is again considered as an unknown quantity. However, the crucial difference from classical inference is that θ is considered to be a *random* variable. Hence, the model-specification $[Y|\theta]$ must be converted to a *joint* distribution for Y and θ by specifying a marginal distribution for θ, hence $[Y, \theta] = [Y|\theta][\theta]$. The marginal distribution of θ is also called the *prior* for θ. Its role is to describe the (lack of) knowledge about θ in the absence of the data, Y. The process of inference then consists of asking how conditioning on the realised data, y, changes the prior for θ into its corresponding *posterior* distribution, $[\theta|y]$. The mechanics of this are provided by Bayes' Theorem,

$$[\theta|Y] = [Y|\theta][\theta]/[Y],$$

where $[Y] = \int [Y|\theta][\theta]d\theta$ is the marginal distribution of Y induced by the combination of the specified model, or likelihood function, and the specified prior.

Bayesian inferential statements about θ are expressed as probabilities calculated from the posterior, $[\theta|Y]$. For example, the Bayesian counterpart of a β-level confidence set is a β-level *credible set*, defined as any set $\mathcal{S}$ such that

$$P(\theta \in \mathcal{S}|Y) = \beta.$$

If a point estimate of θ is required, candidates include the mean or mode of the posterior distribution. Operationally, the crucial requirement for Bayesian inference is the evaluation of the integral which gives the marginal distribution of Y. For many years, this requirement restricted the practical application of Bayesian inference to simple problems. For complex problems and data structures, classical inference involving numerical evaluation and maximisation of the likelihood function was a more practical strategy. However, the situation changed radically with the recognition that Monte Carlo methods of integration, and in particular Markov chain Monte Carlo methods of the kind proposed in Hastings (1970), could be used to generate simulated samples from the posteriors in very complex models. As a result, Bayesian methods are now used in many different areas of application.

A.4   Prediction

We now compare classical and Bayesian approaches to prediction. To do so, we need to expand our model specification, $[Y|\theta]$, to include a *target* for prediction, T, which is another random variable. Hence, the model becomes $[T, Y|\theta]$, a specification of the joint distribution of T and Y for a given value of θ. From a classical inferential perspective, we then need to manipulate the model using Bayes' Theorem to obtain the *predictive* distribution for T as the corresponding conditional, $[T|Y, \theta]$. The data give us the realised value of Y, and to complete the predictive inference for T we can either plug-in the maximum likelihood estimate $\hat{\theta}$ or examine how the predictive distribution varies over a range of values of θ determined by its confidence set.

From a Bayesian perspective, the relevant predictive distribution is $[T|Y]$ i.e., the distribution of the target conditional on what has been observed. Using standard conditional probability arguments, we can express this as

$$[T|Y] = \int [T, \theta|Y] d\theta$$
$$= \int [T|Y, \theta][\theta|Y] d\theta,$$

which shows that the Bayesian predictive distribution is a weighted average of plug-in predictive distributions, with the weights determined by the posteror for θ.

Note that under either the classical plug-in or the Bayesian approach, the answer to a prediction question is a probability distribution. If we want to summarise this distribution, so as to give a *point prediction*, an obvious candidate summary is the mean i.e., the conditional expectation of T given Y. As discussed in Chapter 2, a theoretical justification for this choice is that it mimimises mean square prediction error. However, we emphasise that in general, the mean is just one of several reasonable summaries of the predictive distribution.

Notice that if θ has a known value, then the manipulations needed for classical prediction of T are exactly the manipulations needed for Bayesian inference treating T as a parameter. It follows that in the Bayesian approach, in which parameters are treated as random variables, the distinction between estimation and prediction is not sharp; from a strictly mathematical point of view, the two are identical. Nevertheless, we feel that it is useful to maintain the distinction to emphasise that estimation and prediction address different scientific questions.

References

Azzalini, A. (1996). *Statistical Inference: Based on the Likelihood*, Chapman and Hall, London.

Baddeley, A. and Vedel Jensen, E. B. (2005). *Stereology for Statisticians*, Chapman and Hall/CRC, Boca Raton.

Banerjee, S. (2005). On geodetic distance computations in spatial modeling, *Biometrics* **61**: 617–625.

Banerjee, S., Wall, M. M. and Carlin, B. P. (2003). Frailty modelling for spatially correlated survival data, with application to infant mortality in Minnesota, *Biostatistics* **4**: 123–142.

Barry, J., Crowder, M. and Diggle, P. J. (1997). Parametric estimation using the variogram, *Technical Report ST-97-06*, Dept. Maths and Stats, Lancaster University, Lancaster, UK.

Bartlett, M. S. (1955). *Stochastic Processes*, Cambridge University Press.

Bartlett, M. S. (1964). A note on spatial pattern, *Biometrics* **20**: 891–892.

Bartlett, M. S. (1967). Inference and stochastic process, *Journal of the Royal Statistical Society, Series A* **130**: 457–478.

Bellhouse, D. R. (1977). Some optimal designs for sampling in two dimensions, *Biometrika* **64**: 605–611.

Ben-jamma, F., Marino, M. and Loaiciga, H. (1995). Sampling design for contaminant distribution in lake sediments, *Journal of Water Resources Planning and Managmeent* **121**: 71–79.

Benes, V., Bodlak, K., Møller, J. and Waagepetersen, R. P. (2001). Bayesian analysis of log gaussian cox process models for disease mapping, *Technical Report Research Report R-02-2001*, Department of Mathematical Sciences, Aalborg University.

Berger, J. O., De Oliveira, V. and Sansó, B. (2001). Objective Bayesian analysis of spatially correlated data, *Journal of the American Statistical Association* **96**: 1361–1374.

Besag, J. E. (1974). Spatial interaction and the statistical analysis of lattice systems (with discussion), *Journal of the Royal Statistical Society, Series B* **36**: 192–225.

Besag, J. and Mondal, D. (2005). First-order intrinsic autoregressions and the de Wijs process, *Biometrika* **92**: 909–920.

Boussinesq, M., Gardon, J., Kamgno, J., Pion, S. D., Gardon-Wendel, N. and Chippaux, J. P. (2001). Relationships between the prevalence and intensity of *loa loa* infection in the Central province of Cameroon, *Annals of Tropical Medicine and Parasitology* **95**: 495–507.

Bowman, A. W. and Azzalini, A. (1997). *Applied Smoothing Techniques for Data Analysis*, Oxford University Press.

Box, G. E. P. and Cox, D. R. (1964). An analysis of transformations (with discussion), *Journal of the Royal Statistical Society, Series B* **26**: 211–252.

Breslow, N. E. and Clayton, D. G. (1993). Approximate inference in generalized linear mixed models, *Journal of the American Statistical Association* **88**: 9–25.

Brix, A. and Diggle, P. J. (2001). Spatio-temporal prediction for log-Gaussian Cox processes, *Journal of the Royal Statistical Society, Series B* **63**: 823–841.

Brix, A. and Møller, J. (2001). Space-time multitype log Gaussian Cox processes with a view to modelling weed data, *Scandinavian Journal of Statistics* **28**: 471–488.

Capeche, C. L. e. (1997). Caracterização pedológica da fazenda angra - pesagro/rio - estação experimental de campos (rj), *Informação, globalização, uso do solo*, Vol. 26, Congresso Brasileiro de Ciência do Solo, Embrapa/SBCS, Rio de Janeiro.

Chilès, J.-P. and Delfiner, P. (1999). *Geostatistics: Modeling Spatial Uncertainty*, Wiley, New York.

Christensen, O. (2001). *Methodology and applications in non-linear model based geostatistics*, PhD thesis, Aalborg University, Denmark.

Christensen, O. F. (2004). Monte Carlo maximum likelihood in model-based geostatistics, *Journal of Computational and Graphical Statistics* **13**: 702–718.

Christensen, O. F., Møller, J. and Waagepetersen, R. P. (2001). Geometric ergodicity of Metropolis-Hastings algorithms for conditional simulation in generalised linear mixed models, *Methodology and Computing in Applied Probability* **3**: 309–327.

Christensen, O. F. and Ribeiro Jr., P. J. (2002). geoRglm: a package for generalised linear spatial models, *R-NEWS* pp. 26–28.
*http://cran.R-project.org/doc/Rnews

Christensen, O. F., Roberts, G. O. and Skøld, M. (2006). Robust Markov chain Monte Carlo methods for spatial generalized linear mixed models, *Journal of Computational and Graphical Statistics* **15**: 1–17.

Christensen, O. F. and Waagepetersen, R. P. (2002). Bayesian prediction of spatial count data using generalized linear mixed models, *Biometrics* **58**: 280–286.

Cleveland, W. S. (1979). Robust locally weighted regression and smoothing scatterplots, *Journal of the American Statistical Association* **74**: 829–836.

Cleveland, W. S. (1981). Lowess: A program for smoothing scatterplots by robust locally weighted regression, *American Statistician* **35**: 54.

Cochran, W. G. (1977). *Sampling Techniques*, second edn, Wiley, New York.

Cox, D. R. (1955). Some statistical methods related with series of events (with discussion), *Journal of the Royal Statistical Society, Series B* **17**: 129–157.

Cox, D. R. (1972). Regression models and life tables (with discussion), *Journal of the Royal Statistical Society, Series B* **34**: 187–220.

Cox, D. R. and Hinkley, D. V. (1974). *Theoretical Statistics*, Chapman and Hall, London.

Cox, D. R. and Miller, H. D. (1965). *The Theory of Stochastic Processes*, Methuen, London.

Cox, D. R. and Oakes, D. (1984). *Analysis of Survival Data*, Chapman and Hall, London.

Cressie, N. (1985). Fitting variogram models by weighted least squares, *Mathematical Geology* **17**: 563–586.

Cressie, N. (1993). *Statistics for Spatial Data*, Wiley, New York.

Cressie, N. and Hawkins, D. M. (1980). Robust estimation of the variogram, *Mathematical Geology* **12**: 115–125.

Cressie, N. and Wikle, C. K. (1998). The variance-based cross-variogram: you can add apples and oranges, *Mathematical Geology* **30**: 789–799.

Dalgaard, P. (2002). *Introductory Statistics with R*, Springer.

Davis, J. C. (1972). *Statistics and Data Analysis in Geology*, second edn, Wiley, New York.

De Oliveira, V., Kedem, B. and Short, D. A. (1997). Bayesian prediction of transformed Gaussian random fields, *Journal of the American Statistical Association* **92**: 1422–1433.

De Wijs, H. J. (1951). Statistics of ore distribution. Part I. Frequency distribution of assay values, *Journal of the Royal Netherlands Geological and Mining Society* **13**: 365–375.

De Wijs, H. J. (1953). Statistics of ore distribution. Part II. Theory of binomial distributions applied to sampling and engineering problems, *Journal of the Royal Netherlands Geological and Mining Society* **15**: 12–24.

Diggle, P. J. (2003). *Statistical Analysis of Spatial Point Patterns*, second edn, Edward Arnold, London.

Diggle, P. J., Harper, L. and Simon, S. (1997). Geostatistical analysis of residual contamination from nuclear weapons testing, *in* V. Barnett and F. Turkman (eds), *Statistics for the Environment 3: pollution assessment and control*, Wiley, Chichester, pp. 89–107.

Diggle, P. J., Heagerty, P., Liang, K. Y. and Zeger, S. L. (2002). *Analysis of Longitudinal Data*, second edn, Oxford University Press, Oxford.

Diggle, P. J. and Lophaven, S. (2006). Bayesian geostatistical design, *Scandinavian Journal of Statistics* **33**: 55–64.

Diggle, P. J., Moyeed, R. A., Rowlingson, B. and Thomson, M. (2002). Childhood malaria in the Gambia: a case-study in model-based geostatistics, *Applied Statistics* **51**: 493–506.

Diggle, P. J., Ribeiro Jr, P. J. and Christensen, O. F. (2003). An introduction to model-based geostatistics, *in* J. Møller (ed.), *Spatial Statistics and Computational Methods*, Springer, pp. 43–86.

Diggle, P. J., Tawn, J. A. and Moyeed, R. A. (1998). Model based geostatistics (with discussion), *Applied Statistics* **47**: 299–350.

Diggle, P. J., Thomson, M. C., Christensen, O. F., Rowlingson, B., Obsomer, V., Gardon, J., Wanji, S., Takougang, I., Enyong, P., Kamgno, J., Remme, J., Boussinesq, M. and Molyneux, D. H. (2006). Spatial modeling and prediction of *loa loa* risk: decision making under uncertainty, *International Journal of Epidemiology (submitted)* .

Diggle, P., Rowlingson, B. and Su, T. (2005). Point process methodology for on-line spatio-temporal disease surveillance, *Environmetrics* **16**: 423–434.

Draper, N. and Smith, H. (1981). *Applied Regression Analysis*, second edn, Wiley, New York.

Dubois, G. (1998). Spatial interpolation comparison 97: foreword and introduction, *Journal of Geographic Information and Decision Analysis* **2**: 1–10.

Duchon, J. (1977). Splines minimising rotation-invariant semi-norms in Sobolev spaces, *in* W. Schempp and K. Zeller (eds), *Constructive Theory of Functions of Several Variables*, Springer, pp. 85–100.

Fedorov, V. V. (1989). Kriging and other estimators of spatial field characteristics, *Atmospheric Environment* **23**: 175–184.

Gelfand, A. E., Schmidt, A. M., Banerjee, S. and Sirmans, C. F. (2004). Nonstationary multivariate process modeling through spatially varying coregionalization (with discussion), *Test* **13**: 263–312.

Gelman, A., Carlin, J. B., Stern, H. S. and Rubin, D. B. (2003). *Bayesian Data Analysis*, second edn, Chapman and Hall, London.

Geyer, C. J. (1992). Practical Markov chain Monte Carlo (with discussion), *Statistical Science* **7**: 473–511.

Geyer, C. J. (1994). On the convergence of Monte Carlo maximum likelihood calculations, *Journal of the Royal Statistical Society, Series B* **56**: 261–274.

Geyer, C. J. and Thompson, E. A. (1992). Constrained Monte Carlo maximum likelihood for dependent data (with discussion), *Journal of the Royal Statistical Society, Series B* **54**: 657–699.

Gilks, W. R., Richardson, S. and Spiegelhalter, D. J. (eds) (1996). *Markov Chain Monte Carlo in Practice*, Chapman and Hall, London.

Gneiting, T. (1997). *Symmetric Positive Definite Functions with Applications in Spatial Statistics*, PhD thesis, University of Bayreuth.

Gneiting, T., Sasvári, Z. and Schlather, M. (2001). Analogues and correspondences between variograms and covariance functions, *Advances in Applied Probability* **33**.

Gotway, C. A. and Stroup, W. W. (1997). A generalized linear model approach to spatial data analysis and prediction, *Journal of Agricultural, Biological and Environmental Statistics* **2**: 157–178.

Greig-Smith, P. (1952). The use of random and contiguous quadrats in the study of the structure of plant communities, *Annals of Botany* **16**: 293–316.

Guttorp, P., Meiring, W. and Sampson, P. D. (1994). A space-time analysis of of ground-level ozone data, *Environmetrics* **5**: 241–254.

Guttorp, P. and Sampson, P. D. (1994). Methods for estimating heterogeneous spatial covariance functions with environmental applications, *in* G. P. Patil and C. R. Rao (eds), *Handbook of Statistics X11: Environmental Statistics*, Elsevier/North Holland, New York, pp. 663–690.

Handcock, M. S. and Wallis, J. R. (1994). An approach to statistical spatial temporal modeling of meteorological fields (with discussion), *Journal of the American Statistical Association* **89**: 368–390.

Handcock, M. and Stein, M. (1993). A Bayesian analysis of kriging, *Technometrics* **35**: 403–410.

Harville, D. A. (1974). Bayesian inference for variance components using only error contrasts, *Biometrika* **61**: 383–385.

Hastie, T. (1996). Pseudosplines, *Journal of the Royal Statistical Society, Series B* **58**: 379–396.

Hastings, W. K. (1970). Monte Carlo sampling methods using Markov chains and their applications, *Biometrika* **57**: 97–109.

Henderson, R., Shimakura, S. E. and Gorst, D. (2002). Modelling spatial variation in leukaemia survival data, *Journal of the American Statistical Association* **97**: 965–972.

Higdon, D. (1998). A process-convolution approach to modelling temperatures in the North Atlantic ocean (with discussion), *Environmental and Ecological Statistics* **5**: 173–190.

Higdon, D. (2002). Space and space-time modelling using process convolutions, *in* C. Anderson, V. Barnett, P. C. Chatwin and A. H. El-Shaarawi (eds), *Quantitative Methods for Current Environmental Issues*, Wiley, Chichester, pp. 37–56.

Hougaard, P. (2000). *Analysis of Multivariate Survival Data*, Springer.

Journel, A. G. and Huijbregts, C. J. (1978). *Mining Geostatistics*, Academic Press, London.

Kammann, E. E. and Wand, M. P. (2003). Geoadditive models, *Applied Statistics* **52**: 1–18.

Kent, J. T. (1989). Continuity properties of random fields, *Annals of Probability* **17**: 1432–1440.

Kitanidis, P. K. (1978). Parameter uncertainty in estimation of spatial functions: Bayesian analysis, *Water Resources Research* **22**: 499–507.

Kitanidis, P. K. (1983). Statistical estimation of polynomial generalized covariance functions and hydrological applications, *Water Resources Research* **22**: 499–507.

Knorr-Held, L. and Best, N. (2001). A shared component model for detecting joint and selective clustering of two diseases, *Journal of the Royal Statistical Society, Series A* **164**: 73–85.

Kolmogorov, A. N. (1941). Interploation und extrapolation von stationären zufälligen folgen, *Izv. Akad. Nauk SSSR* **5**: 3–14.

Krige, D. G. (1951). A statistical approach to some basic mine valuation problems on the Witwatersrand, *Journal of the Chemical, Metallurgical and Mining Society of South Africa* **52**: 119–139.

Lark, R. M. (2002). Optimized spatial sampling of soil for estimation of the variogram by maximum likelihood, *Geoderma* **105**: 49–80.

Laslett, G. M. (1994). Kriging and splines: an empirical comparison of their predictive performance in some applications, *Journal of the American Statistical Association* **89**: 391–409.

Lee, Y. and Nelder, J. A. (1996). Hierarchical generalized linear models (with discussion), *Journal of the Royal Statistical Society, Series B* **58**: 619–678.

Lee, Y. and Nelder, J. A. (2001). Modelling and analying correlated non-normal data., *Statistical Modelling* **1**: 3–16.

Li, Y. and Ryan, L. (2002). Modelling spatial survival data using semiparametric frailty models, *Biometrics* **58**: 287–297.

Liang, K. Y. and Zeger, S. L. (1986). Longitudinal data analysis using generalized linear models, *Biometrika* **73**: 13–22.

Mardia, K. V. and Watkins, A. J. (1989). On multimodality of the likelihood in the spatial linear model, *Biometrika* **76**: 289–296.

Matérn, B. (1960). Spatial Variation, *Technical report*, Statens Skogsforsningsinstitut, Stockholm.

Matérn, B. (1986). *Spatial Variation*, second edn, Springer, Berlin.

Matheron, G. (1963). Principles of geostatistics, *Economic Geology* **58**: 1246–1266.

Matheron, G. (1971a). Random set theory and its application to stereology., *Journal of Microscopy* **95**: 15–23.

Matheron, G. (1971b). The theory of regionalized variables and its applications, *Technical Report 5*, Cahiers du Centre de Morphologie Mathematique.

Matheron, G. (1973). The intrinsic random functions and their applications, *Advances in Applied Probability* **5**: 508–541.

McBratney, A. B. and Webster, R. (1981). The design of optimal sampling schemes for local estimation and mapping of regionalised variables. II. Program and examples., *Computers and Geosciences* **7**: 335–365.

McBratney, A. B. and Webster, R. (1986). Choosing functions for semi-variograms of soil properties and fitting them to sample estimates, *Journal of Soil Science* **37**: 617–639.

McBratney, A., Webster, R. and Burgess, T. (1981). The design of optimal sampling schemes for local estimation and mapping of regionalised variables. I. Theory and methods., *Computers and Geosciences* **7**: 331–334.

McCullagh, P. and Nelder, J. A. (1989). *Generalized Linear Models*, second edn, Chapman and Hall, London.

Menezes, R. (2005). Assessing spatial dependency under non-standard sampling. Unpublished Ph.D. thesis.

Metropolis, N., Rosenbluth, A. W., Rosenbluth, M. N., Teller, A. H. and Teller, E. (1953). Equations of state calculations by fast computing machine, *Journal of Chemical Physics* **21**: 1087–1091.

Møller, J., Syversveen, A. R. and Waagepetersen, R. P. (1998). Log-Gaussian Cox processes, *Scandinavian Journal of Statistics* **25**: 451–482.

Møller, J. and Waagepetersen, R. P. (2004). *Statistical Inference and Simulation for Spatial Point Processes*, Chapman and Hall/CRC.

Muller, W. (1999). Least squares fitting from the variogram cloud, *Statistics and Probability Letters* **43**: 93–98.

Muller, W. G. and Zimmerman, D. L. (1999). Optimal designs for variogram estimation, *Environmetrics* **10**: 23–27.

Natarajan, R. and Kass, R. E. (2000). Bayesian methods for generalized linear mixed models, *Journal of the American Statistical Association* **95**: 222–37.

Naus, J. I. (1965). Clustering of random points in two dimensions, *Biometrika* **52**: 263–267.

Neal, P. and Roberts, G. O. (2006). Optimal scaling for partially updating MCMC algorithms, *Annals of Applied Probability* **16**: 475–515.

Nelder, J. A. and Wedderburn, R. M. (1972). Generalized linear models., *Journal of the Royal Statistical Society, Series A* **135**: 370–84.

O'Hagan, A. (1994). *Bayesian Inference*, Vol. 2b of *Kendall's Advanced Theory of Statistics*, Edward Arnold.

Omre, H. (1987). Bayesian kriging — merging observations and qualified guesses in kriging, *Mathematical Geology* **19**: 25–38.

Omre, H., Halvorsen, B. and Berteig, V. (1989). A Bayesian approach to kriging, *in* M. Armstrong (ed.), *Geostatistics*, Vol. I, pp. 109–126.

Omre, H. and Halvorsen, K. B. (1989). The Bayesian bridge between simple and universal kriging, *Mathematical Geology* **21**: 767–786.

Patterson, H. D. and Thompson, R. (1971). Recovery of inter-block information when block sizes are unequal, *Biometrika* **58**: 545–554.

Pawitan, Y. (2001). *In All Likelihood: Statistical Modelling and Inference Using Likelihood*, Oxford University Press, Oxford.

Perrin, O. and Meiring, W. (1999). Identifiability for non-stationary spatial structure, *Journal of Applied Probability* **36**: 1244–1250.

R Development Core Team (2005). *R: A language and environment for statistical computing*, R Foundation for Statistical Computing, Vienna, Austria. http://www.R-project.org.

Rathbun, S. L. (1996). Estimation of poisson intensity using partially observed concomitant variables, *Biometrics* **52**: 226–242.

Rathbun, S. L. (1998). Spatial modelling in irregularly shaped regions: kriging estuaries, *Environmetrics* **9**: 109–129.

Ripley, B. D. (1977). Modelling spatial patterns (with discussion), *Journal of the Royal Statistical Society, Series B* **39**: 172–192.

Ripley, B. D. (1981). *Spatial Statistics*, Wiley, New York.

Ripley, B. D. (1987). *Stochastic Simulation*, Wiley, New York.

Ross, S. (1976). *A First Course in Probability*, Macmillan, New York.

Royle, J. A. and Nychka, D. (1988). An algorithm for the construction of spatial coverage designs with implementation in splus, *Computers and Geosciences* **24**: 479–88.

Rue, H. and Held, L. (2005). *Gaussan Markov Random Fields: Theory and Applications*, Chapman and Hall, London.

Rue, H. and Tjelmeland, H. (2002). Fitting Gaussian random fields to Gaussian fields, *Scandinavian Journal of Statistics* **29**: 31–50.

Ruppert, D., Wand, M. P. and Carroll, R. J. (2003). *Semiparametric Regression*, Cambridge University Press, Cambridge.

Russo, D. (1984). Design of an optimal sampling network for estimating the variogram, *Soil Science Society of America Journal* **52**: 708–716.

Sampson, P. D. and Guttorp, P. (1992). Nonparametric estimation of nonstationary spatial covariance structure, *Journal of the American Statistical Association* **87**: 108–119.

Sarndal, C. E. (1978). Design-based and model-based inference in survey sampling (with discussion), *Scandinavian Journal of Statistics* **5**: 27–52.

Schlather, M. (1999). Introduction to positive definite functions and to unconditional simulation of random fields, *Technical Report ST-99-10*, Dept. Maths and Stats, Lancaster University, Lancaster, UK.

Schlather, M., Ribeiro Jr, P. J. and Diggle, P. J. (2004). Detecting dependence between marks and locations of marked point processes, *Journal of the Royal Statistical Society, Series B* **66**: 79–93.

Schmidt, A. M. and Gelfand, A. E. (2003). A bayesian corregionalization approach for multivariate pollutant data, *Journal of Geophysical Research — Atmospheres* **108 (D24)**: 8783.

Schmidt, A. M. and O'Hagan, A. (2003). Bayesian inference for nonstationary spatial covariance structures via spatial deformations, *Journal of the Royal Statistical Society, Series B* **65**: 743–758.

Serra, J. (1980). Boolean model and random sets, *Computer Graphics and Image Processing* **12**: 99–126.

Serra, J. (1982). *Image Analysis and Mathematical Morphology*, Academic Press, London.

Spruill, T. B. and Candela, L. (1990). Two approaches to design of monitoring networks, *Ground Water* **28**: 430–442.

Stein, M. L. (1999). *Interpolation of Spatial Data: Some Theory for Kriging*, Springer, New York.

Takougang, I.and Meremikwu, M., Wanji, S., Yenshu, E. V., Aripko, B., Lamlenn, S., Eka, B. L., Enyong, P., Meli, J., Kale, O. and Remme, J. H. (2002). Rapid assessment method for prevalence and intensity of *loa loa* infection, *Bulletin of the World Health Organisation* **80**: 852–858.

Tanner, M. (1996). *Tools for Statistical Inference*, Springer, New York.

Thomson, M. C., Connor, S. J., D'Alessandro, U., Rowlingson, B. S., Diggle, P. J., Cresswell, M. and Greenwood, B. M. (1999). Predicting malaria infection in Gambian children from satellite data and bednet use surveys: the importance of spatial correlation in the interpretation of results, *American Journal of Tropical Medicine and Hygiene* **61**: 2–8.

Thomson, M. C., Obsomer, V., Kamgno, J., Gardon, J., Wanji, S., Takougang, I., Enyong, P., Remme, J. H., Molyneux, D. H. and Boussinesq, M. (2004). Mapping the distribution of *loa loa* in cameroon in support of the african programme for onchocerciasis control, *Filaria Journal* **3**: 7.

Van Groenigen, J. W., Pieters, G. and Stein, A. (2000). Optimizing spatial sampling for multivariate contamination in urban areas, *Environmetrics* **11**: 227–244.

Van Groenigen, J. W., Siderius, W. and Stein, A. (1999). Constrained optimisation of soil sampling for minimisation of the kriging variance, *Geoderma* **87**: 239–259.

Van Groenigen, J. W. and Stein, A. (1998). Constrained optimisation of spatial sampling using continuous simulated annealing, *Journal of Environmental Quality* **27**: 1076–1086.

Wahba, G. (1990). *Spline Models for Observational Data*, Society for Industrial and Applied Mathematics.

Waller, L. A. and Gotway, C. A. (2004). *Applied Spatial Statistics for Public Health Data*, Wiley, New York.

Warnes, J. J. and Ripley, B. D. (1987). Problems with likelihood estimation of covariance functions of spatial Gaussian processes, *Biometrika* **74**: 640–642.

Warrick, A. and Myers, D. (1987). Optimization of sampling locations for variogram calculations, *Water Resources Research* **23**: 496–500.

Watson, G. S. (1971). Trend-surface analyis, *Mathematical Geology* **3**: 215–226.

Watson, G. S. (1972). Trend surface analyis and spatial correlation, *Geology Society of America Special Paper* **146**: 39–46.

Wedderburn, R. W. M. (1974). Quasilikelihood functions, generalized linear models and the Gauss-Newton method, *Biometrika* **63**: 27–32.

Whittle, P. (1954). On stationary processes in the plane, *Biometrika* **41**: 434–449.

Whittle, P. (1962). Topographic correlation, power-law covariance functions, and diffusion, *Biometrika* **49**: 305–314.

Whittle, P. (1963). Stochastic processes in several dimensions, *Bulletin of the International Statistical Institute* **40**: 974–974.

Winkels, H. and Stein, A. (1997). Optimal cost-effective sampling for monitoring and dredging of contaminated sediments, *Journal of Environmental Quality* **26**: 933–946.

Wood, A. T. A. and Chan, G. (1994). Simulation of stationary Gaussian processes in $[0, 1]^d$, *Journal of Computational and Graphical Statistics* **3**: 409–432.

Wood, S. N. (2003). Thin plate regression splines, *Journal of the Royal Statistical Society B* **65**: 95–114.

Zhang, H. (2002). On estimation and prediction for spatial generalized linear mixed models, *Biometrics* **58**: 129–136.

Zhang, H. (2004). Inconsistent estimation and asymptotically equal interpolations in model-based geostatistics, *Journal of the American Statistical Association* **99**: 250–261.

Zimmerman, D. L. (1989). Computationally efficient restricted maximum likelihood estimation of generalized covariance functions, *Mathematical Geology* **21**: 655–672.

Zimmerman, D. L. and Homer, K. E. (1991). A network design criterion for estimating selected attributes of the semivariogram, *Environmetrics* **4**: 425–441.

Zimmerman, D. L. and Zimmerman, M. B. (1991). A comparison of spatial semivariogram estimators and corresponding kriging predictors, *Technometrics* **33**: 77–91.

Index

Binomial logit-linear model, 14, 82-83, 94-95

Continuity/differentiability
  mean square, 49-51
  path continuity, 49-51
Correlation function
  Matérn, 29, 51-53
  powered exponential, 52-54
  spherical, 55
  wave, 55-56
Cox process, 86-87

Data
  Gambia malaria, 4-5, 24-25, 26, 184-186
  Loa loa prevalence,186-193
  Paraná , 26, 45
  Rongelap island, 2-4, 22-24, 26, 180-184, 196
  salinity monitoring, 205-207
  soil, 5-8, 44-45, 120-122, 127-128, 150-151
  surface elevations, 1-2, 17-20, 26, 30-39, 43, 113-115, 131, 166-168
  Swiss rainfall, 117-120, 148-149, 168-171,
Design, 12, 27-28, 197-210
  Bayesian, 203-209

lattice plus close pairs, 200-201
lattice plus in-fill, 200-201
prospective, 207-209
retrospective, 204-207
uniform, 199-201
Directional effects, 57-60
Distance, definitions of, 39

Gaussian model, 13, 46-77
  intrinsic, 62-66
  linear, 14
  low-rank, 68-69
  multivariate, 69-74
  stationary, 13, 29
  transformed, 60-63, 116-117, 165-166
Generalized estimating equations, 124-125
Generalized linear model, 13-15, 78-97, 96 97
Geostatistical model, 9

Hierarchical likelihood, 124
Hypothesis testing, 13

Inference
  Bayesian
  classical

Kriging

ordinary, 136-137
simple, 136-137, 139-145
trans-Gaussian, 146-148
with non-constant mean, 38-39, 150

Low-rank models, 68-69

Markov chain Monte Carlo, 159,
        171-172, 176-178
Monte Carlo maximum likelihood, 123
Multivariate methods, 10-11, 69-74

Nugget effect, 56-57

Parameter estimation, 12, 15-16
    Bayesian 156-166, 172-175, 213-214
    least squares, 106-111, 129-130
    maximum likelihood, 115-116,
        122-123, 130-131, 211-213
    restricted maximum likelihood,
        111-115
    weighted least squares, 107-109
Parameter estimation and prediction,
    distinction between, 35-36
Point process models, 85-88
Poisson log-linear model, 14, 81-82,
    93-94
Prediction, 12, 15-16, 133-155, 214-215
    Bayesian, 157-159, 175-176
    minimum mean square error, 133-148
    plug-in, 156-157
    with nugget effect, 138-139
Preferential sampling, 12, 88-92, 95-96
Profile likelihood, 113

Random sets, 93
Regularisation, 48-49

Sample locations, choice of, 199-201
Scan processes, 93
Signal, 13
Simulation
    conditional, 67-68
    unconditional, 66-68
Software, 17-25
    geoR, 16-25, 41-44, 74-76,93-96,
        125-131, 150-154, 192-195
    geoRglm, 16-17, 195
    GMRFlib, 75
    Random Fields, 75
    R-project, 16
Support, 9-10
Survival analysis, 83-85

Transformation, 30
Trend
    estimation of, 99
    external, 11, 57
    surface, 11, 57

Variogram
    bivariate
    directional, 103-104
    empirical, 33-35, 101-103, 125-127
    of generalised linear model, 80-81
    theoretical, 46-48, 99-101

Springer Series in Statistics *(continued from p. ii)*

Lahiri: Resampling Methods for Dependent Data.
Le/Zidek: Statistical Analysis of Environmental Space-Time Processes.
Le Cam: Asymptotic Methods in Statistical Decision Theory.
Le Cam/Yang: Asymptotics in Statistics: Some Basic Concepts, 2nd edition.
Liu: Monte Carlo Strategies in Scientific Computing.
Manski: Partial Identification of Probability Distributions.
Mielke/Berry: Permutation Methods: A Distance Function Approach.
Molenberghs/Verbeke: Models for Discrete Longitudinal Data.
Mukerjee/Wu: A Modern Theory of Factorial Designs.
Nelsen: An Introduction to Copulas, 2nd edition.
Pan/Fang: Growth Curve Models and Statistical Diagnostics.
Politis/Romano/Wolf: Subsampling.
Ramsay/Silverman: Applied Functional Data Analysis: Methods and Case Studies.
Ramsay/Silverman: Functional Data Analysis, 2nd edition.
Reinsel: Elements of Multivariate Time Series Analysis, 2nd edition.
Rosenbaum: Observational Studies, 2nd edition.
Rosenblatt: Gaussian and Non-Gaussian Linear Time Series and Random Fields.
Särndall/Swensson/Wretman: Model Assisted Survey Sampling.
Santner/Williams/Notz: The Design and Analysis of Computer Experiments.
Schervish: Theory of Statistics.
Shaked/Shanthikumar: Stochastic Orders.
Shao/Tu: The Jackknife and Bootstrap.
Simonoff: Smoothing Methods in Statistics.
Sprott: Statistical Inference in Science.
Stein: Interpolation of Spatial Data: Some Theory for Kriging.
Taniguchi/Kakizawa: Asymptotic Theory for Statistical Inference for Time Series.
Tanner: Tools for Statistical Inference: Methods for the Exploration of Posterior
    Distributions and Likelihood Functions, 3rd edition.
Tillé: Sampling Algorithms.
Tsaitis: Semiparametric Theory and Missing Data.
van der Laan/Robins: Unified Methods for Censored Longitudinal Data and
    Causality.
van der Vaart/Wellner: Weak Convergence and Empirical Processes: With
    Applications to Statistics.
Verbeke/Molenberghs: Linear Mixed Models for Longitudinal Data.
Weerahandi: Exact Statistical Methods for Data Analysis.

Statistical Analysis of Environmental Space-Time Processes

Nhu D. Le and James V. Zidek (Editors)
The book focuses on environmental space-time fields and, to a more limited extent, the assessment of the risks they pose to human health in particular. The book showcases Bayesian and empirical Bayesian theory for assessing and tackling risks, such as spatial prediction, cross-validatory assessment, monitoring, and measurement error, with emphasis on methods developed by the authors. Difficult conceptual issues are addressed as well as the problems they pose for consultants, subject area investigators, and statistical scientists. Solutions are described in non-technical language with intuitive motivation, to provide an operational understanding for those who require sophisticated tools for their analysis of environmental fields.

2006. 352 p. (Springer Series in Statistics) Hardcover ISBN 0-387-26209-1

Space, Structure and Randomness
Contributions in Honor of Georges Matheron in the Fields of Geostatistics, Random Sets, and Mathematical Morphology

Michel Bilodeau, Fernand Meyer and Michel Schmitt (Editors)
This volume is divided in three sections on random sets, geostatistics and mathematical morphology. They reflect his professional interests and his search for underlying unity. Some readers may be surprised to find theoretical chapters mixed with applied ones. GM always considered that the distinction between the theory and practice was purely academic.

2005. 402 p. (Lecture Notes in Statistics) Softcover ISBN 0-387-20331-1

Case Studies in Spatial Point Process Modeling

A. Baddeley, P. Gregori, J. Mateu, R. Stoica, and D. Stoyan (Editors)
Point process statistics is successfully used in fields such as material science, human epidemiology, social sciences, animal epidemiology, biology, and seismology. Its further application depends greatly on good software and instructive case studies that show the way to successful work. This book satisfies this need by a presentation of the spatstat package and many statistical examples.

2006. 306 p. (Lecture Notes in Statistics) Softcover ISBN 0-387-28311-0